D1008348

THE ORION NEBULA

C. ROBERT O'DELL

The Belknap Press of

Harvard University Press

Cambridge, Massachusetts

London, England 2003

The ORION NEBULA

Where Stars Are Born

Printed in the United States of America

Designed by Gwen Frankfeldt

Library of Congress Cataloging-in-Publication Data

O'Dell, C. Robert (Charles Robert), 1937–
The Orion Nebula : where stars are born /
C. Robert O'Dell.
p. cm.
Includes index.
ISBN 0-674-01183-X (alk. paper)
1. Orion Nebula. I. Title.

QB855.9.O75O34 2003
523.1'13—dc21
2003050332

*

CONTENTS

PREFACE

NOTHING surpasses the night sky viewed from an isolated mountaintop. The winter sky is special because it is dominated by the constellation Orion. Not only is this region rich in lore, but it is also the home of one of the most famous objects in the sky, the Orion Nebula.

More than fifty years ago, I stared at Orion as an amateur astronomer working with a telescope I had assembled from parts, then later with ones I had constructed "from scratch." I've since spent a lifetime building telescopes and instrumentation. This hardware was simply a means to the end of learning more about how the universe works. The Orion Nebula, always an object of fascination for me, was a natural target as better observational equipment was developed. That first small telescope I used to examine the glowing cloud in Orion's sword has been succeeded by the world's most powerful telescopes, including the Keck Telescope in Hawaii and the Hubble Space Telescope. The Orion Nebula always seems to reserve a surprise for each new instrument.

In this book I share today's knowledge of the Orion Nebula with readers who are interested in astronomy, but who may lack a strong technical background. To really appreciate the nebula as something more than a thing of beauty, you will need to understand a few basic processes. These I explain in Chapters 1–7 and 9. This short course on the night sky and how it has been viewed and interpreted is presented historically, linking revolutionary ideas and improved understanding to pioneer astronomers and astrophysicists, while never losing sight of our target, the Orion Nebula.

As project scientist for the Hubble Space Telescope, I felt a sense of continuity with my professional ancestors as we built the world's most powerful observatory. In Chapter 10 I give a unique insider's view of the origin of this observatory and its construction, including the causes of the flaws present when it was launched, and in Chapter 11 I discuss how this revolutionary telescope has revealed a new view of the Orion Nebula.

The point of a better telescope, of course, is a fuller understanding of what you're looking at. Not only is the Orion Nebula a thing of beauty, it is also our closest rich cradle for the formation of stars and planets like our own. When we study the Orion Nebula of today, we are looking at conditions similar to our origins 5 billion years ago.

THE ORION NEBULA

CHAPTER ONE

Enter the Hunter

THE murmurs in a large telescope's dome are a mighty roar when compared to the silence of the stars. But we measure our world by our own experiences. As a young astronomer using the 120-inch telescope of the Lick Observatory, near San Francisco Bay, I was having the usual problems staying alert throughout a nearly silent, clear night. In 1963 remote control of giant telescopes had yet to be developed, so that I was perched inside the top end of the instrument as it pointed around the sky. Mounting the telescope was best done in the dark since it involved a vertigo-inducing step from the dome's ladder into the cramped quarters of the observer's cage. My night quickly settled into a routine of pointing at various targets in our Milky Way Galaxy. Tonight these were mostly the ill-named planetary nebulae. Planetary nebulae are not planets—rather, they are the outer layers of stars like our Sun thrown off as these stars collapse from having exhausted all of their nuclear fuel. Over my shoulder, I glimpsed the easily identified constellation Orion the Hunter, the three bright stars of his belt rising in the east. Orion was the prime target for the second half of the night. These three bright stars are actually massive young stars, outshining our Sun 100,000-fold, but reduced to bright pinpoints by their distance of 1,500 light years. The light I saw from those stars, in other words, took 1,500 years to reach my eyes. My target for the late night was not in Orion's Belt, however, but in a vertical grouping of stars that form the Hunter's sword, in particular the central region of this grouping that appears as a smudge, rather than a crisp star, to the unaided eye. This is the Orion Nebula.

In this earlier age of stellar observation, telescopes could not be pointed accurately. There were no sensitive TV cameras to pipe images down to a heated control room. These improvements arrived during the flowering of the physical sciences that followed the launch of the Soviet spacecraft *Sputnik* in 1957. Astronomers had to actually look through the telescope to identify their target and align its image onto the particular scientific instrument being used. With the Lick Telescope this was accomplished by swinging the prime focus spectrograph, which spread light into a spectrum and recorded an image of that spectrum photographically on a glass plate, out of the way and using an eyepiece for examining the image formed by the giant mirror sitting fifty feet away. For the typically myopic scientist there was the ability to disregard the eyepiece and examine the image directly with two lenses, one the great mirror and the other in one's eye.

Nothing in my experience had prepared me for what swept into view as the telescope moved quickly to catch the rising hunter's sword. I had seen many pictures of the region and had begun my apprenticeship in a circular graduate student's office that featured the most detailed drawing ever made of the object. However, the unexpected view was a show-stopping image. At the center were four bright blue stars—so bright that they almost overwhelmed the hundreds of other fainter stars surrounding them. From extremely faint to very bright, the clustering of stars was one of the richest in the sky. Their colors varied smoothly, with the brightest being the blue tint of those four stars and the faintest being deep red rubies sprinkled across the picture. But the most striking thing was the glowing cloud, the nebula. It filled my view, being brightest not at the brightest stars, but off-center, with a bright barred feature and many small clumps. The colors were sublime, but clear. The central region had a blue-green glow, which gradually changed to red away from the central stars, while the bright bar was a clear red. If you're not impressed by the Orion Nebula, it's unlikely you'll find much to impress you at all in the night sky.

I was so impressed that I have spent much of my professional life studying the Orion Nebula, using new observational techniques and analytical tools as they became available; most recently the Hubble Space Telescope, arguably the most powerful telescope of our day. In this book I will explain what we know about the Orion Nebula and how we derived this knowledge, and provide a brief background in astronomy that will help place this famous object in its proper context.

Orion the Hunter is the constellation that dominates the sky in the early evening in December, as shown in Figure 1.1. It is different from most constellations in that the majority of its stars are actually associated with one another, lying at similar distances. The three stars that form Orion's belt are easily picked out, being close together and among the brightest stars in the sky. Hanging below the belt stars are three more stars forming Orion's sword, much fainter than those in the belt, but still easily visible with the naked eye. The conjunction of keen eyes, a new Moon, and no city lights will further reveal to an observer that the middle star of these three is actually a bit fuzzy. This is the Orion Nebula. To the ancients, this middle star was the eighth brightest star in the constellation, hence it bears the name Theta Orionis, since the brightest stars were designated according to their brightness—the brightest being Alpha, the second Beta, and so on.

In Central America, the Mayas have a folk tale associated with this region of the sky. Their traditional hearths are composed of three stones, which they identify with the two lowest bright stars in the constellation, together with the leftmost (east on the sky) of the belt stars. In the middle of these celestial hearthstones is Theta Orionis, a smudge of glowing fire. This association is important for telling the early history of the Orion Nebula because it is clear pretelescopic evidence that the ancients saw something diffuse in the sky there, whereas most objects are crisp. It is one matter for modern observers who know that the nebula is there to claim that they can see it with the unaided eye, but it is a better authentication that the object was thought to be a fuzzy star long before telescopes could confirm the fact.

The puzzling fact is that the man noted as the father of the astronomi-

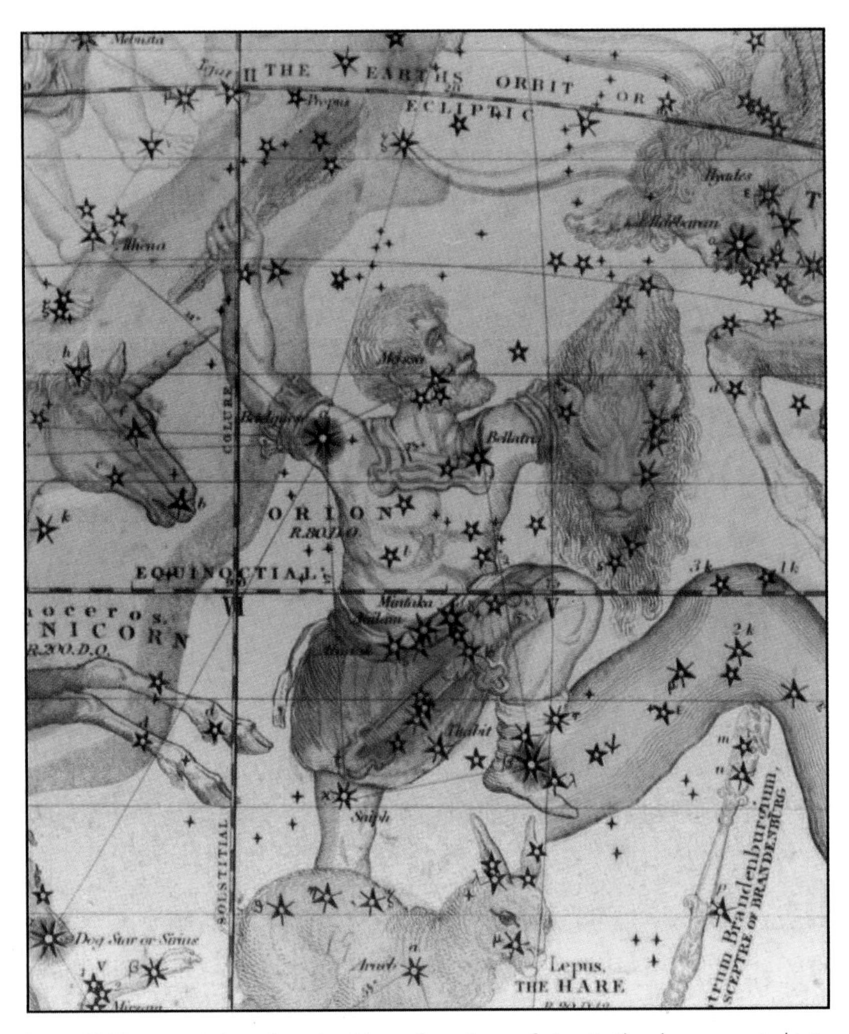

Figure 1.1 The association of myth with configurations of stars in the sky was carried into the nineteenth century, by which time the constellation Orion was nearly two millennia old. Orion is the most striking constellation in the early winter evening sky in the United States. The three bright stars in the belt of the hunter are easy to recognize. The Orion Nebula is associated with the center of the grouping of three stars in the middle of the sheathed sword (F. J. Huntington celestial map of 1835, detail, collection of the author).

cal telescope, Galileo Galilei, did not record seeing the Orion Nebula. I cautiously designate him as the father of the astronomical telescope, a title that he well deserves, although certainly he was not the inventor of the telescope. By 1609 lenses had been in use for several centuries for eyeglasses to aid the near- and far-sighted. Making a lens is a relatively simple matter and doesn't require understanding the laws of physics that govern the formation of a lens. It is good enough to know that if you make a lens one particular way it helps near-sightedness and if made another way it helps far-sightedness. Galileo learned that in the Netherlands it had been found that a particular combination of lenses of two distinctive shapes (one convex on both sides, the other concave on both sides) would form a magnified image of a distant object when used together. Galileo immediately identified the utility of such a device for military purposes and quickly sought funding for his increasingly refined telescopes (things aren't much different four centuries later). When he turned his telescopes to the sky he made a dazzling series of discoveries—craters on the Moon, spots on the Sun, satellites going around Jupiter, and so on. Even though he thoroughly mapped the sword stars in Orion, he did not note a nebula there. Probably the reason for this failure was the poor quality of the glass he was using, which had imbedded grit and bubbles. These imperfections made all bright stars appear hazy, so that the bright stars in the core of Theta Orionis would not have looked much different from other stars, even though they had an intrinsic cloud of emission associated with them.

Nicholas Peiresc, an Italian astronomer then working in Marseille, France, is usually credited with the discovery of the Orion Nebula in 1610 with a telescope of the style of Galileo's and no better capability. However, the report was only in his own records, so that when the Jesuit astronomer Johann Baptist Cysat mentioned it in a book on a bright comet in 1618, he gained arguable credit as the discoverer. However, his description—"Another [example] of this phenomenon in the heavens is the congeries of stars in the last [sic] star in the sword of Orion, for there one can distinguish a similar congestion of some stars in a very narrow space, and all around and between the stars themselves is a diffused light like a radiant white cloud"—is rather ambiguous. Often discoveries are credited to the person who firmly establishes both the presence and detail of an object or phenomenon, and also indicates its importance. If this is the case, then the credit is due Christiaan Huygens, who published a drawing of the nebula in his book, *Systema Saturnium,* in 1659 (Figure 1.2). The modern nebula is clearly recognizable in this depiction, with Theta Orionis divided into multiple stars, presaging today's knowledge that Theta Orionis is actually a rich cluster of stars associated with the Orion Nebula.

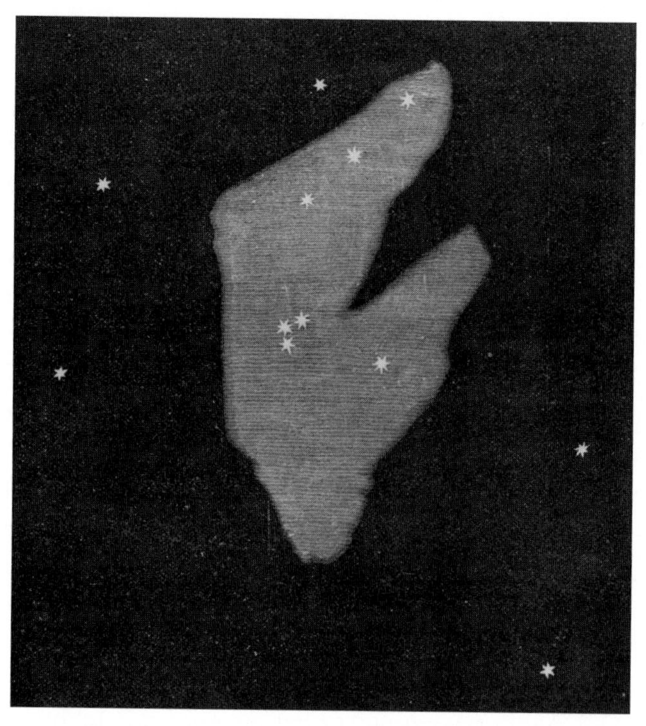

Figure 1.2 This drawing by Christiaan Huygens, published in 1659, is thought to be the first drawing of the Orion Nebula. Note that the drawing is oriented so that north points toward the lower right corner (Owen Gingerich, Harvard College Observatory).

Throughout the next century the Orion Nebula remained a favorite target of astronomers. Images became progressively better as larger telescopes were manufactured. Usually this took the form of making one's own instrument. It is much easier to make the optics for a telescope with a long focal length. The focal length is the distance between the main lens and where the image is formed. The image created is then examined by an eyepiece. A telescope with a short focal length is harder to make. Unless a longer focal length is accompanied by a larger aperture (the diameter of the main light-gathering component) of the lens, however, the light of a nebula is spread out and the object is harder to see. This meant that the gigantic telescopes built for the study of the Moon and the planets weren't very good for the Orion Nebula. As techniques improved, telescopes began to have larger lenses in proportion to their focal lengths, so that fuzzy objects in the sky started getting more attention. Of course the most famous fuzzy objects are comets, which inspired particular awe because they were simply not understood, even after Edmund Halley predicted the reappearance of one comet (Halley's comet) in 1758, which proved to be after his death. Comet hunting was a competitive scientific sport in the eighteenth century, just as it is now among amateur astronomers. However, comets often look a lot like other immobile fuzzy objects in the sky, such as nebulae. Charles Messier of France compiled a list of some hundred similar objects that were not comets as an aid to comet seekers. Spread out across the sky, these objects were listed by number: the Orion Nebula was Messier 42 (usually shortened to M 42). His list of objects includes some of the most spectacular objects in the sky, such as the Crab Nebula (M 1) and the great nebula in Andromeda (M 31). He also drew M 42 in greater detail than ever before (Figure 1.3).

The German-born English astronomer William Herschel started building larger and larger reflecting telescopes in the late eighteenth century from a bronze alloy known as speculum metal. This allowed an image-forming (by reflection) surface to be generated with much less effort than earlier refracting telescopes, which had a double lens up front, each with two surfaces. Moreover, the reflector telescope lenses (mirrors) were

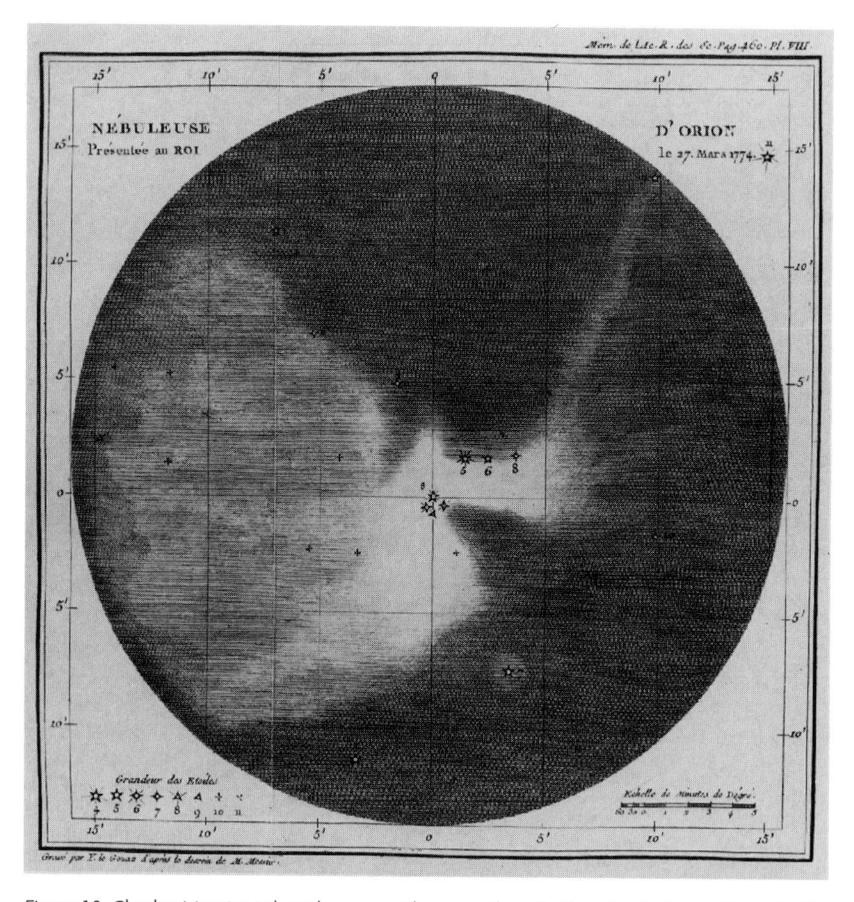

Figure 1.3 Charles Messier is best known as the compiler of a list of nebulous objects in the sky that might be confused with comets. He was also an astute observer and in 1771 published this drawing of the Orion Nebula. North is at the bottom of the figure, a standard orientation until the early twentieth century. The four stars grouped in the center are called the Trapezium (Owen Gingerich, Harvard College Observatory).

easier to keep rigid, since they could be supported on their back sides. The largest of the speculum mirror telescopes was that of William Parsons, the Earl of Rosse in English-controlled Ireland, whose "Leviathan of Parsonstown" had an aperture of six feet. The mirror was so massive that its mounting allowed it to observe objects only as they passed a slice of the sky aligned exactly north-south. The great apertures of the Herschel and Parsons telescopes helped enormously in observing faint stars, but it was already apparent that the quality of the image was really the most important factor for observing bright objects like the Orion Nebula. As we'll see in Chapter 5, a well-built telescope of only five inches aperture produces about the best-quality image the turbulent conditions of Earth's atmosphere normally allow. Thus it is hardly surprising that the best ground-based telescope drawing of the Orion Nebula is that of George Bond, who made it using the Harvard 15-inch refractor during periods of particularly stable conditions over four years (Figure 1.4).

Figure 1.4 George Bond's drawing of the Orion Nebula, the result of several years of careful visual study of the region, was published in 1877. It is usually considered the most accurate drawing of the nebula, but was superseded by photographic images only a few years later. North is at the top of the figure (Owen Gingerich, Harvard College Observatory).

The era of illustrating the Orion Nebula came to an abrupt halt in September 1880 when Henry Draper made the first photograph of this object. Although the photographic emulsions of the day were not particularly good and telescope stability was poor, it became clear that the future of Orion images and astonomical images in general would be photographic, at least for the next century or more.

C H A P T E R T W O

Views of Our Universe

IN THE case of astronomical photos, a bit of background information only adds to our sense of wonder and appreciation of the night sky's beauty. Images of the Orion Nebula are no exception to this rule—their complexity can be overwhelming. For this reason I'll spend the next several chapters explaining the hows and the whys that are the intellectual background to these lovely pictures. If you are really impatient, jump ahead to Chapter 8, then come back.

The Orion Nebula is both an object and a process, the process of turning the debris from dying stars into a new generation of stars, most very much like our own Sun. These transformations have been occurring all over the universe since quite early in its 13-billion-year history, but to narrow the field I will concentrate on our home galaxy, the box within which our world was created.

The observer fortunate enough to find a dark sky for viewing the heavens will be familiar with the Milky Way. Galileo used his first telescopes to resolve this seemingly cloudlike band of light into individual stars. Eventually it was discovered that this band of light is in fact an edge-on view of our home galaxy, the Milky Way. It is a rather garden-variety galaxy, even though it contains more than 10 billion stars. The sheer magnitude of that number is put into perspective when one recognizes that the observable universe contains at least 40 billion galaxies (and many more that are too faint to be seen). We live in a universe of galaxies and ours is just one of them. This plethora of galaxies is the result of fragmentation of the material of the universe, which in its earliest period after the big bang was a primordial soup. The soup wasn't quite homogeneous, however, which means that at only a small fraction of its age, the universe began to fragment, creating the largest structures known, the cosmic voids and shells. Soon after fragmentation into even smaller units began, one of which would become our galaxy.

The most important property of galaxies is that they are gravitationally bound. By that, I mean that the gravitational pull of one part on any other part is enough to keep it from flying off into space. This has the extremely important consequence that each individual galaxy develops and changes independent of all the other galaxies, except for the rather rare incidents when they collide with one another. This is why we can look upon our galaxy as being our continent, essentially cut off from the others, developing in its own way and upon its own timescale. The laws of physics apply to all nascent galaxies, as these laws apply to continents on Earth, but some may develop primates and others marsupials. What actually develops depends on the properties of the cards each incipient galaxy is dealt. Those with lots of mass will become rich in stars, while those without much mass are doomed to being dwarf galaxies. The known galaxies vary over a factor of 10,000 in their mass and even more in their subsequent luminosities. Mass is a relatively simple concept; it is a measure of how much material is there. The other key ingredient for galaxies is more esoteric. It is called angular momentum (to be more precise, angular momentum per unit of mass). Analogous to ordinary momentum, which keeps an object moving ahead, angular momentum is what keeps an object spinning. Some nascent galaxies have a lot of angular momentum, others do not. As a result, some galaxies develop into highly symmetric flattened objects called spiral galaxies, while others will always be highly irregular in form. If the object has little angular momentum per unit mass but lots of total mass, it will develop into a symmetric but not flattened galaxy called an elliptical galaxy. Figure 2.1 illustrates the wide variety of forms that galaxies can take.

Our home is a spiral galaxy. Although much of its mass is in a central bulge, most of the material is in a flattened disk. Our Sun formed in this disk about 5 billion years ago and is located about two-thirds of the way

Figure 2.1 This Hubble Space Telescope image shows some of the wide variety of forms assumed by galaxies. What are not illustrated here are the dwarf galaxies, which are much fainter than anything depicted, the irregular galaxies, and the quasars (the Space Telescope Science Institute and NASA).

out to our galaxy's edge. Like most of the local material, the gravity of this disk holds us very close to the middle of the disk. This means that when we look "up" or "down" we see few stars. When we look along the disk (remember, from inside it), however, we see so many stars that to the unaided eye they blend together into that band sweeping across the sky called the Milky Way. The Milky Way is not the same everywhere, since the number of stars per unit volume increases as one gets closer to the center of our galaxy. This means that when we look toward the center, the Milky Way looks much brighter. This is why the June Milky Way is much easier to observe than the December Milky Way: in June we are pointed toward the center of our galaxy at night, whereas in December we are pointed toward the center during the day. The difference would be even greater if we could see all the way to the center; however, we'll see that this is not possible, at least in terms of visual light. Since we are in the midst of this forest of stars, it is easiest to see the form of our galaxy by looking at images of similar nearby objects, the closest being our neighbor, the Andromeda Galaxy, M 31. You can see this object with the unaided eye, if you know where to look and are in Earth's northern hemisphere. If you observe from the southern hemisphere, then M 31 isn't available; however, two smaller but easier to find neighbor galaxies are. They bear the name of the Magellanic Clouds, reflecting the fact that the survivors of the first round-the-world expedition brought word of these cloudlike objects to Europe. Although we cannot view our galaxy from afar, the overwhelming evidence is that it is a typical large spiral galaxy and would look very much like M 51, which is illustrated in Figure 2.2.

Until less than a hundred years ago, astronomers believed that our galaxy encompassed the entire universe and that our Sun was located near its center. It was as if the anthropocentric view of Ptolemy had simply been replaced with one dealing with greater distances and scales, but retaining man at its center. The Ptolemaic system was a logically self-consistent model of the universe with Earth at the center and the Sun, planets, and stars moving about it. This model explained all of the features that were observed at the time of its formulation in the second century—it could explain the apparent motion of the Sun, Moon, and planets, and the daily motion of the stars. The bulk of the universe was thought to consist of the Sun, Moon, and planets, and the stars were relegated to lights on a shell that rotated about Earth every 24 hours. This explained all the known properties of the sky and better observations of the motions of the planets necessitated only slight alterations to the sys-

Figure 2.2 This Hubble image is of the center of the nearby spiral galaxy M 51. If we were able to move above the plane of our galaxy and look down, it would look very similar to this. Note how the spiral structure extends all the way to the small nucleus. Dark lanes of interstellar matter trace each spiral arm, indicating concentrations of this gas and dust. The red gems along the spiral arms are objects like the Orion Nebula, but vastly larger. The brightest stars are foreground objects in our galaxy (the Space Telescope Science Institute and NASA).

tem. For cultures with a man-centered, monotheistic religion, placing Earth at the center seemed fully appropriate. An alternative model where Earth and other planets moved around the Sun was proposed early in the flourishing of Greek culture, but the Ptolemaic model was not challenged by what we now know to be basically the correct model until the publication in 1543 of Nicolas Copernicus' book, *On the Revolutions.* I call it "basically correct" because Copernicus represented the motions of the planets as circles, whereas in fact they are ellipses. Indeed, the model was no better at explaining the motion of the planets than the Ptolemaic model and the movement of Earth was not confirmed until early in the nineteenth century. However, our Sun-centered model of the solar system received wider acceptance once Galileo and other early telescopic observers of the early seventeenth century saw how there were moons moving around other planets and that the innermost planets underwent phases, like our Moon. Johannes Kepler used the newfound precise measurements of the motion of the planets made by the Danish nobleman Tycho Brahe to show that the Copernican model worked wonderfully well if the planetary motions were ellipses about the Sun. Then in the early eighteenth century Isaac Newton provided the explanation for how it was that planets stayed in orbit around the Sun through his universal law of gravitation. Now the motions of planets, their moons, and the comets could be thought of as components of a clockwork mechanism that ran without our help. However, the nature of the stars and the diffuse objects in the sky remained a mystery. Even the nearest stars are at distances so great that they appeared to be painted on a celestial globe that rotated about Earth every 24 hours (23 hours 56 minutes actually) owing to Earth spinning about its tilted axis (thereby changing the direction of a north-south plane). By the eighteenth century it was accepted that the stars must be highly luminous objects like our Sun, but located at incredibly great distances. It was not clear if the difference in brightness of various stars could be attributed to their different intrinsic luminosities or their greater or lesser distances from us.

As a light source is moved farther away it appears fainter, even though its intrinsic brightness (its absolute luminosity, to use a term employed in the rest of this book) remains the same. There is a precise way of calculating the variation of the apparent luminosity as a function of distance. A series of transparent and concentric spheres surrounding a light source will all have the same amount of light passing through them. However, the amount of light received per unit area will drop as the square (a value multiplied by itself) of the radius of the distance to the source. The apparent brightness will drop as the square of the distance from the light. One can turn this problem around if one knows the absolute luminosity of the source. In this case, one can calculate the distance to the light by measuring its apparent brightness.

Entering from stage right, near the end of the eighteenth century, is the next hero of our story, William Herschel. Trained as a musician, he left his native Germany for England, where the royal family was German, since the bloodline had been saved by calling on a distant relative in Germany, who became George I. A journeyman performer and composer (his music is still available, although more as a novelty than on its absolute merit), Herschel's passion was astronomy. He began observing the heavens with telescopes of his own construction. His speculum metal mirror telescopes were the most powerful of their day. With the assistance of his sister Caroline he scanned the heavens and by chance found a small disk-shaped object. It was bigger than a star and didn't have the fuzzy appearance of a comet. Its motion revealed it to be a planet, the first to be discovered with a telescope. The other six, including Earth, were visible to the naked eye. What a discovery this was, to have found a new world! Herschel had the political acumen to attempt to name it after the king. The scientific establishment prevailed and it became known as Uranus, but his efforts did gain him the favor and financial support of the royal house, so that he could concentrate on making and using new telescopes.

One of his numerous projects was to use his telescopes as a depth gauge for studying how stars are distributed through space. To achieve this, he counted the stars of various apparent brightness with his telescope, doing this in multiple directions throughout the sky. He then made the reasonable assumption that stars all have the same intrinsic luminosity as our Sun and calculated the distance of the faintest stars accordingly. What he found is illustrated in Figure 2.3. We now understand

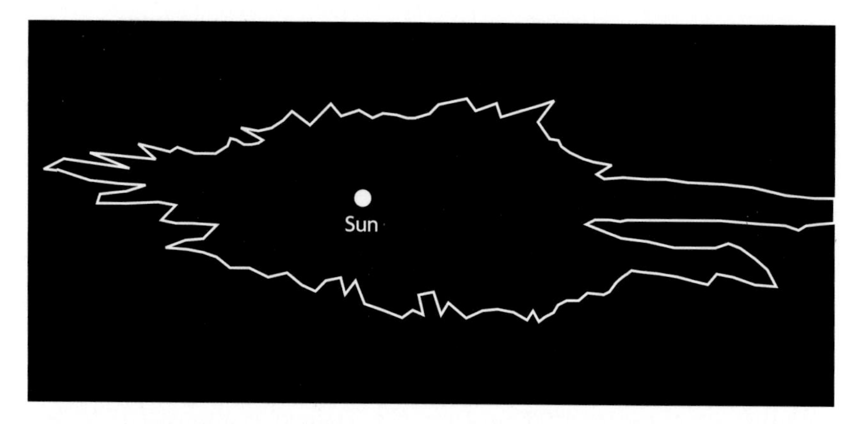

Figure 2.3 Sir William Herschel attempted to model a cross-section of our galaxy by counting the number of stars down to a specific brightness level in various directions in and above the bright band of the Milky Way. This figure shows the model he derived. We now know that interstellar material dims and obscures more distant stars, so that this model largely reflects the effects of this extinction rather than the true distribution of stars. Nonetheless, this basic model survived as the best model of our galaxy for a century. Herschel thought the total size was only a few thousand light years.

his model to be simply a section of the local part of our galaxy, but Herschel thought that our galaxy was the entire universe. This means that he had derived a model of the universe. Looking along a line across the sky passing through the middle of the Milky Way (the galactic equator), he found the stars he believed to be farthest away. His size was slightly larger in the direction we now know to be the center of our galaxy. Perpendicular to the galactic equator the stars didn't extend as far. The galaxy was flattened. The farthest part of this universe was about 3,000 light years. A light year is the distance light travels in one year, an incredible 5.9 trillion miles (or 9.5 trillion kilometers). The light year is 63,241 times the average distance between the center of the Sun and the center of Earth (a term called the Astronomical Unit, or AU). Herschel's development of a quantitative model of our galaxy, and hence the universe, was a tremendous achievement, one which was refined over the next century. Unfortunately, it was entirely wrong.

Although Herschel's method was basically sound, it contained a fatal flaw. It turns out that stars have a wide range of absolute luminosity, but this doesn't make any difference if one averages over a large sample of stars, because their average comes out to be about that of our Sun. The reason Herschel's model and those derived from it were incorrect is that he was not aware that the space in between the stars was not empty. In the same way that intervening haze will cause a light to become fainter than it would if the light were only diminished by distance, a haze of fine dust particles in between the stars was making them fainter than they should have been. If fact, the farther one goes, the greater the effect, so that when Herschel looked out into the plane of our galaxy, he could only see stars for a limited distance. There were stars beyond, but he couldn't observe them because of this "interstellar extinction." His estimation of the distribution of stars perpendicular to the Milky Way was reasonably accurate because most of the stars in that direction are nearby, and there is, therefore, little extinction. This is the thin dimension of the disk that is our galaxy.

Herschel's model of the universe was vastly larger than the Ptolemaic model. Again we appeared to be at the center of the universe, a position not uncomfortable if you were taught that the universe was created for your benefit.

In his defense, Herschel did posit that our galaxy was not the entire universe. He had discovered and drawn a wealth of nebulae (clouds) that had no obviously associated stars and he proposed that these were actually gravitationally bound systems of stars (which they are) observed at such great distances that the individual stars are not seen. However, Herschel's theory was not proven until 1924, when the remarkably successful young astronomer, Edwin Powell Hubble, established the nebulae's great distances. We therefore live in a universe of island universes, to mix terms.

Once it was recognized that our galaxy is but one of many galaxies and that our views of stars must be corrected for the effects of interstellar extinction, astronomers developed and refined an accurate model throughout the remainder of the twentieth century. This refinement was expedited in large part by the advent of radio telescopes, which can see radiation that is not effected by the interstellar haze. In this modern picture the center of our galaxy is about 25,000 ly away, the site of an in-

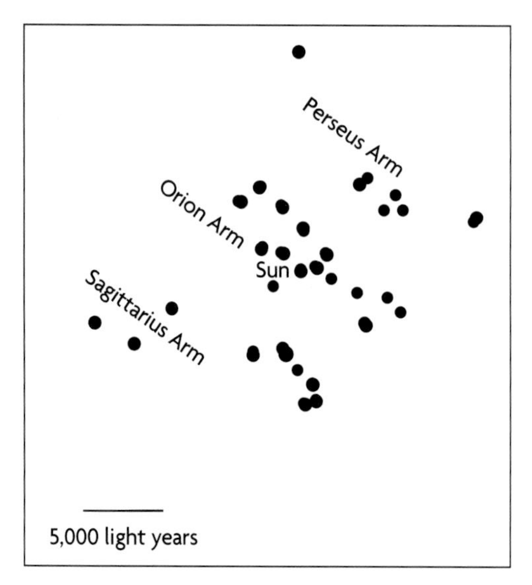

Figure 2.4 This map of the location of clusters of bright young stars near the Sun shows how they delineate spiral structure, much like taking a local sample of the galaxy M 51 shown in Figure 2.2.

tense gravitational field, which may be caused by a massive black hole, a phenomenon invoked to explain the enormous absolute luminosity of some galaxies. At a lesser level it is frequently found in otherwise normally appearing galaxies. We also know that most of the material is spread in a disk composed of stars, interstellar dust, and interstellar gas. This disk rotates about the center of our galaxy, but not like the solid rotation of a CD, on which two specks of dust always have the same position relative to one another. The disk of our galaxy is in differential rotation, that is, the rotation period is different for different distances from the center, with the inner parts making a revolution about the center in less time than the outer. This is very similar to what is happening in our solar system, where the innermost planets have much shorter "years" than the outer planets. Superimposed on this galactic disk of material are elongated features known as spiral arms, whose counterparts are easy to see in our neighboring spiral galaxies. The spiral arms of our galaxy are not obvious to us because we look out from a position inside the disk and the spiral arms overlap one another on the sky. The existence of these spiral arms is very important for understanding the Orion Nebula, because they contain the greatest concentrations of interstellar dust and gas, essential ingredients for star formation. Our Sun formed in such a spiral arm, but after the twenty-five revolutions that it has made since then, we must be well removed from the point of origin and our parent spiral arm has probably long since disappeared and been replaced by others.

If we could look down on our local section of the galaxy, we would see something that looks like Figure 2.4. Our Sun is located in a region outside of any spiral arm, but there is one lying close to us. The stars forming the constellation Orion mostly lie in this spiral arm and provide its name. The region depicted here is about the same size as Herschel's model of the "universe." It is not a universe, of course, but only a local neighborhood within a rather ordinary galaxy among the billions of such objects. It is important to us because it is where multiple generations of stars have formed, where the centers of these stars have created the heavy elements that make life possible—in short, because it is our own. From it we have a clear view of the continuing process of star formation occurring in the Orion Nebula.

C H A P T E R   T H R E E

Henry Draper and the Photographic Revolution

SCIENTISTS like facts and irrefutable evidence, free of personal bias. The introduction of photography into the practice of astronomy represented a tremendous advance in the consistency and reliability of astronomical data. Before photography astronomers had to rely upon a written or illustrated record of the night sky. There was ample room for inadvertent bias. Although an obviously hearty lot, judging by tales of astronomers hiking up to mountaintops in the afternoon, then observing all night, they were often operating in the cold and dark and late at night. These are far from ideal conditions for making and recording observations. Few had the advantage of William Herschel, whose sister Caroline shared his intellectual and professional life. Caroline recorded what William saw through the telescope, thus avoiding that draconian decision to lose precious observing time to record what was being seen. Most kept their eye trained on the telescope, relying on memory to record their facts later, always a risky process. When the observation involved a transitory phenomenon (for example, seeing a new moon of a planet when it was at just enough distance to be seen) or fell at the edge of the astronomer's ability to detect something, then the old system demonstrated its shortcomings. Nineteenth-century astronomy journals are filled with often highly personal exchanges of letters claiming and disputing observations.

Not only were these observations being made by persons with a variety of skills and levels of objectivity, there was also the natural tendency to fit something new into an existing paradigm. This is particularly true when seeing the unexpected. While at the Yerkes Observatory of the University of Chicago early in my career, I and my colleagues would almost always receive UFO reports by telephone whenever Venus was bright and high in the evening sky. Seeing the unexpected, callers had simply jumped to a conclusion, attributing to the unknown object the UFO characteristics they'd read about—flying around, then disappearing. Such a report was even turned in by an engineer who later became president of the United States!

I have made a similar blunder. While observing at the University of Wisconsin's Pine Bluff Observatory, I had just closed the telescope dome and was walking away in the predawn glow. In the east and near where the Sun was about to appear I saw a totally new and quite bright object. I stood there studying it, recording the time and mentally fixing just what I was seeing—a highly curved glowing arc—then rushed back and opened the telescope. When I had the telescope pointed there, there was nothing. The object had looked very much like a comet. I had been observing comets while serving as an assistant to my thesis advisor and knew that comets were much brighter when closer to the Sun and were usually discovered in the glow of dawn or dusk. Imagine my surprise when the local paper carried a photograph the next day of the results of shooting a rocket up over Lake Erie, then releasing its payload of highly reflective radar chaff (flakes of aluminum foil) into the predawn sky. For a brief period the chaff glowed from the sunlight that was hitting it at its high altitude and the cloud was easy to see against the night sky. From where I stood in southern Wisconsin, the cloud of chaff was near the rising point of the Sun. The picture, taken from a nearby point by the military, who were conducting the test, showed that it in no way resembled a comet. However, I had unknowingly categorized the object as comet-like and recorded it as cometary in form. Surely the eye-brain system is a marvelous one, but it has many limitations, including lack of permanence and therefore the inability to re-examine the evidence.

The advent of astronomical photography removed much of the personal element in observation and allowed the detection of vastly fainter objects. Astronomy began to take its modern form as a science, with the facts clearly laid out for examination and debate. Photography provided a permanent record of objects that simply could not be seen otherwise. This technology matured at the same time the nature of science in the United States was changing. Up through the Civil War, science had been the provenance of the gentleman scholar, those whose financial situation gave them the time to pursue intellectual endeavors commensurate with their abilities. There was no National Science Foundation (NSF) distributing funds to the fortunate winners of highly competitive programs. Colleges often had observatories, but they were not often used for what we would today call research, but rather for teaching and enjoyment. Fortunately, the modern university took shape in the United States at just about the same time as astronomical photography began. A series of independent universities was created—Johns Hopkins in 1876, Stanford and Chicago in 1891—each trying to model itself after the famous European universities that had been the center of intellectual pursuit for some time (although the gentleman-scientist was a major player in European science too). Existing colleges expanded and shifted their mission during this same period, creating the modern forms of schools like Harvard and Yale. All of this reflected the growing wealth and strength of the United States. Indeed, universities were often funded by wealthy benefactors as appropriate memorials to themselves. The face of astronomy was forever altered, both by photography and by the evolution of the modern scientific research institution.

The Invention of Photography

We identify the Wright brothers as the inventors of powered aircraft, since it was their monumental achievement, built on the early steps of their predecessors, that produced the leap into controlled, sustained, manned flight. There are no Wright brothers of photography. By that, I mean that there is no person or point in time when photography (writing with light) was invented. Early experimentation in photography involved identifying a substance that was altered by exposure to light and could then have that alteration permanently fixed by a chemical process prior to subsequent exposure to light. There were many candidate processes. If I were to single out a single inventor of a practical process, it would be Nicephore Niepce, working in France in 1826. His work was followed by William Henry Fox Talbot's successes in 1834, those of Louis Daguerre in 1837, and Fox Talbot's calotype process in 1840. The various processes had their own foibles and advantages; some were very insensitive to light and others produced a positive image, making them difficult to reproduce. The famous daguerreotypes were positive images made directly onto a metal plate. The wet collodion plate had been developed by 1851. This was a glass plate (a stable and cheap supporting material) with the light-sensitive materials coated on one surface, offering good resolution of detail and the advantages of being a negative image. However, it was awkward to use because it required treatment of the plate immediately before the exposure, because the exposure had to be made while the plate was still wet. Civil War photographs were made by this difficult process. Wet plates had no real potential for astro-photography outside of the Sun and Moon because only those objects were bright enough to permit the necessary short exposures.

The technical breakthrough that made astro-photography possible was the development of the dry gelatin plate in England in 1871. In this case the photoactive chemicals were suspended in a solution of gelatin and spread in a thin layer on a glass plate, remaining sensitive even after the gelatin had dried. This meant that long exposures could be taken and that the exposed plates did not have to be processed immediately. This was a process that could work at the telescope, although it was developed to satisfy the market need for a more practical and handy photographic process. The biggest step in satisfying that need was taken by George Eastman in 1888 when he developed the ability to suspend the gelatin film on a flexible medium and the roll-film camera became practical.

The First Photograph of the Orion Nebula

The first photographic image of the Orion Nebula was made by Henry Draper, M.D., an amateur, in September, 1880 (see Figure 3.1). Draper was raised in a family where such feats were encouraged. His father, John William Draper, was a professor of chemistry and botany at the New

Figure 3.1 Henry Draper, a wealthy amateur, applied the rapidly evolving technology of photography to astronomy. His list of astronomical firsts includes the first image of the Orion Nebula (Owen Gingerich, Harvard College Observatory).

York University. He was also a founder of the New York Medical School and a pioneer in photography, having been one of the first Americans to master Daguerre's process, and helped to establish a photo studio in 1840. He made the first images of the Moon that showed any lunar features in 1840 and in 1843 made daguerreotypes that showed new features in the solar spectrum. By 1850 John William Draper was making photo-micrographs and drew his teenage son Henry into his projects.

Obviously gifted, Henry Draper was raised in a milieu of scientific innovation, and finished his medical school studies in 1857 at the age of 20. He immediately threw himself into designing and building telescopes. Evidently not hard-pressed financially, he improved his situation further by marrying well and established himself in a fine New York residence and a second home in outlying Dobb's Ferry, which was close to his observatory at Hastings-on-Hudson. He obtained the first stellar spectrum showing absorption lines in 1872 (spectrographs are described in the next chapter). While visiting England in 1879 he learned from the eminent scientist William Huggins of the commercial availability of dry photographic plates even more sensitive than the inconvenient wet plates that Draper was then using. He began making spectra and images of the Moon and bright planets. He then used his 11-inch Clark Brothers photographic refractor on September 30, 1880, to make a fifty-minute exposure of the Orion Nebula. He quickly reproduced and distributed this image widely, without much attempt at interpretation. This first image was succeeded by a much superior image of 137 minutes exposure on March 14, 1882 (see Figure 3.2). He did not have the chance to extensively follow up on this achievement for he died of a respiratory ailment in November 1882 at the age of forty-five. Nonetheless, his achievements and techniques were well understood and built upon by others. It was clear that the photographic process was the wave of the future and visual observations prevailed only in those areas necessitating very brief periods of observation under momentary superb conditions.

Henry Draper has an additional legacy that he never saw. After his death his widow financially supported a program of surveying the 220,000 brightest stars visible from the northern hemisphere and systematically determining their spectral types. This work was carried out at the Harvard College Observatory and was a success because of the efforts of Annie Jump Cannon and her female collaborators. The resulting Henry Draper Catalog opened the field of systematic study of the characteristics of stellar atmospheres, which had its first great triumph in the first decade of the twentieth century, when Eijnar Hertzsprung and Henry Norris Russell independently and correctly ordered stars in the correct spectral sequence and showed how this sequence also indicated a star's intrinsic luminosity. The resulting Hertzsprung-Russell diagram is one of the standard ways of explaining the properties of stars and will be described in more detail in Chapter 6.

The Triumph of Photography

If our eyes were built for viewing really faint objects we would keep running into things. By this I mean that our eye-brain imaging system wipes out the old image and replaces it with a new one about twenty times per second. This characteristic allows us to easily detect moving objects and

Figure 3.2 The first deep photographic image of the Orion Nebula, made by Henry Draper in March 1882, shows more detail than the George Bond drawing published in 1877. North is at the top of the figure. The four Trapezium stars are overexposed in this full image of 137 minutes and lie to the lower right of the Dark Bay feature (Owen Gingerich, Harvard College Observatory).

to see the space in which we move and operate. This wonderful ability to detect motion comes at the price of short exposure times for integrating together the image, which in turn means that we cannot see really faint objects. It is hardly surprising that we have evolved eyes that aren't built for looking through telescopes. It is not that the eye is insensitive—its quantum efficiency approaches one half (i.e., about half of the incoming photons produce a sensible response), a level only recently achieved by modern detectors. The problem is that in one-twentieth of a second we cannot collect many photons. Seeing fainter objects means collecting light for a longer time and we cannot do this with our eyes.

The astronomer can make time exposures with a photographic detector. Henry Draper's first image was made with a photographic emulsion that probably had a quantum efficiency of only 1 percent of 1 percent (i.e., only one in 10,000 photons was recorded). Nevertheless, by exposing 60,000 times longer than the capability of the human eye, he could see to fainter levels. Since 1880 there have been significant advances in making better emulsions. The best are now about 1 percent efficiency, so that only a few seconds' exposure is needed to produce an image as good as you can see. Making a time exposure with an astronomical telescope is not as easy as simply setting your camera on a tripod and leaving the shutter open. Astronomical targets seem to move across the sky as Earth rotates (called diurnal motion). This means that to produce a crisp image you must accurately move the telescope at a rate that exactly compensates for this motion. This technology has existed for two centuries and presents no problem in modern observatories. However, there are limits to how long you can observe an image. The average dark sky lasts only about nine hours per night. In addition, the diurnal motion of a star across the sky means that they usually rise and set (the exceptions are the circumpolar stars). You cannot observe stars near the horizon because Earth's atmosphere absorbs and distorts the image, meaning that you can make good observations only when the objects are at least 30 degrees above the horizon, further limiting the maximum exposure time.

There are other factors that limit exposure time. Long exposures are less efficient, a property called reciprocity failure. At first examination you would think that doubling the exposure time when looking at an object half as bright would produce the same signal, but it doesn't. The reason for this is in the detailed chemistry that occurs in the surface of the emulsion. The image is gradually lost as the emulsion just sits there. If you have the ability to add images together, then it is better to take multiple short images rather than one long one.

Our eyes have the ability to see objects over an enormous range of brightness levels. If it weren't that way our ancestors, strolling across a sunlit meadow, would not have been able to see the predatory wild beast lying in wait under the shade of a tree. Photographic emulsions don't share this property. Technically put, it is said that the photographic emulsion has a limited dynamic range. Any one grain within the emulsion can record only a certain number of photons of light. If you try to add more photons to that grain, it doesn't record them. There are ways to circumvent this problem. The most popular way is to scan each small section of the image and record its transmission, storing the results in a computer. This is called digitization of an image and is a process still in use. Digitization has largely been replaced by using detectors that give a digital image directly, however, as discussed in Chapter 4.

Photography sounds like a difficult process for observing, and it is. When compared for accuracy to written or hand-illustrated observations, however, the balance quickly tips in favor of photography, complicated though it may be. I began this chapter by extolling the virtues of photography over the written record or sketch and I hope that I have made the advantages of photography manifest. However, we must remember that even a permanent photographic record is examined by human observers, who bring a variety of abilities and biases with them to their offices.

CHAPTER FOUR

The Toolbox of the Astronomer

ASTRONOMY differs from other sciences because we cannot directly interact with the subjects we study. The physicist, chemist, biologist, and geologist can all sample their subjects, learn about them by performing tests, and determine how they react to certain environments and how they can be altered. This is not to say that such steps are easy. Physicists may have to build an enormous accelerator in order to cause energetic collisions of the components of an atom's nucleus. Geologists may have to drill deep into Earth's crust to obtain their sample. Biologists may have to painstakingly isolate a rare cell line. At least they can hope to control their experiments. By controlling an experiment they can test a hypothesis or alter a reaction in order to draw insight from the results.

Astronomy, on the other hand, is an observational science. It deals with objects that we cannot touch and thereby influence. We can make observations about their behavior, but we cannot intervene. Although these statements are true for astronomy as a whole, they do not apply to the subdiscipline of solar system science. Once man could directly sample the flow of wind from our Sun, examine a sample of lunar material, or touch the surface of Venus or Mars, those subjects left their traditional spot within the rubric of astronomy and became solar system science.

Outside of solar system science, however, we must content ourselves with observing from afar. Even the tools astronomers use are purely observational. The history of modern astronomy, which began with the development of the optical telescope, has been one of steady progress through improved technology and knowledge of the underlying physics. The rest of this chapter will be given over to describing some of the tools that astronomers use. However, a brief digression is in order to describe an intellectual tool that applies across all of modern science.

The Scientific Method

There is an often misused term bantered about in our language. You may recall that in high school you were drilled on a well-defined set of steps used by scientists: form a hypothesis, perform an experiment, and draw conclusions. There is nothing wrong with this description of the basic method of scientific inquiry, except that it rarely plays out in exactly this way. If you ask active research scientists to describe the scientific method, you'd get a long pause, then perhaps an earful of information about the specifics of their research. In other words, it's hard to describe, but easy to recognize.

Two ingredients common to the scientific method are a respect for facts and a skepticism about all things thought to be facts. Scientists build their ideas and conclusions around factual information, not about what they wish or feel must be true. This respect for facts is balanced by the realization that there is nothing that is an absolute, inviolate fact—nothing that does not contain within it the possibility, however remote, of being disproven. Some things come closer to an absolute fact than others, of course. The fact that Earth rotates with a period of twenty-four hours is about as close to a fact as one can get. Every piece of evidence supports this hypothesis and the predictions of other models that can explain why objects in the sky appear to move simply don't agree with what one observes.

The closer you come to the cutting edge of science the fuzzier the facts become. Determining the age of the universe by measuring the Hubble constant (the constant of proportionality between the distance to a galaxy and its velocity of recession from us) is an excellent example. Three-quarters of a century of trying to pin down the magnitude of this "constant" has seen its accepted value drop by a factor of ten. Specialists still argue for values that differ from each other by as much as 30 percent. An even more recent example of how the facts become uncertain near the cutting edge is the argument that the universe is actually accelerating in its rate of expansion, a conclusion that is only valid if one accepts a long series of results, each with an intrinsic uncertainty. When these uncertainties are compounded, they render the final answer uncertain.

As the facts become fuzzy there is more of an opening for the human factor. Scientists often want things to turn out a certain way, sometimes because of self-ascribed insight, competition with another scientist, laziness, or the desire to produce something new. Good scientists may yield to any (perhaps all) of these, but they never lose sight of the shortcomings of what they are saying. Chutzpah counts, but in the end it will be the open market of competition of ideas and facts that will win the day.

There is an additional force operating in science and that is the tendency to reinforce doctrine. The Earth-centered model of the solar system is the perfect example: As observational facts accumulated and independent arguments were identified the Ptolemaic model was simply refined, adding complexities to the original simple model until it was horribly complex. This detailed model produced satisfactory agreement with the observational facts of the day. Not until Copernicus was able to think outside of the envelope of what had become doctrine did a new model emerge. Kepler's identification of the laws of motion of planets (elliptical orbits about the Sun) produced in one stroke a working variation of Copernicus' Sun-centered model of the solar system. Copernicus' model was intrinsically simple yet fully explained all of the information available. In a similar case, the highly refined Kapteyn model of the universe turned out to be entirely wrong. Science seems to proceed in fits and starts. Sometimes conceptual breakthroughs are the result of the painstaking accrual of information sufficient to establish the inadequacy of the accepted interpretive model, while other times (more irritating to established scientists) the new paradigms arise from insightful dilettantes unencumbered by too many facts.

The Optical Telescope

Modern astronomy employs a vast array of instruments, but most information still comes in from the optical telescope, which has been in use for almost 400 years. Although there are a variety of designs, I'll emphasize the characteristics of large research telescopes. Bigger is almost always better, because the brightness of an image depends upon the area of the aperture. Our eye has a maximum diameter of about a quarter-inch, which means that a 100-inch telescope forms an image 160,000 times brighter than the unaided eye.

The image has to be formed so that it is not degraded by the telescope itself and must be located where something useful can be done with it. This is commonly accomplished by using a large mirror to collect light and form the image. Mirrors have multiple advantages, including the fact that light of all colors is formed into an image at the same spot and the glass that supports the thin reflective surface can be supported from the rear. When the first giant telescope was constructed (the 200-inch or 5-m at Palomar Mountain), auxiliary equipment didn't take up much space and was located in the center of the telescope. Although this equipment blocked some light, its relatively small size made it possible to avoid installing an additional mirror and losing some light from that added reflection. Modern auxiliary equipment is much more efficient and sensitive, but also takes up more space. Additional mirrors are necessary for depositing the image at a location outside of the telescope tube.

An excellent example of a modern telescope is the Keck 10-m telescope on Mauna Kea in Hawaii (actually there is a pair of them), which is shown in Figure 4.1. In this telescope the image is formed at a position that is highly stable. It can support great weight, and even large instruments do not block out additional light. The primary mirror is a good example of how telescope designs have changed. Before the Keck, telescopes were made with monolithic mirrors, i.e., mirrors made of a single piece of material (usually glass or something similar because it is both stable and much lighter than metal). However, the weight of a monolithic mirror of a given strength goes up as the cubed power (a number multiplied by itself two times) of its aperture. This is why no effective telescopes bigger than the 200-inch were built in the fifty years after construction of the Palomar giant. However, the designers of the Keck circumvented that problem by making the mirror out of thirty-six individual mirrors, then holding them into alignment with one another by very precise position-sensing devices. The result is a mirror weighing but

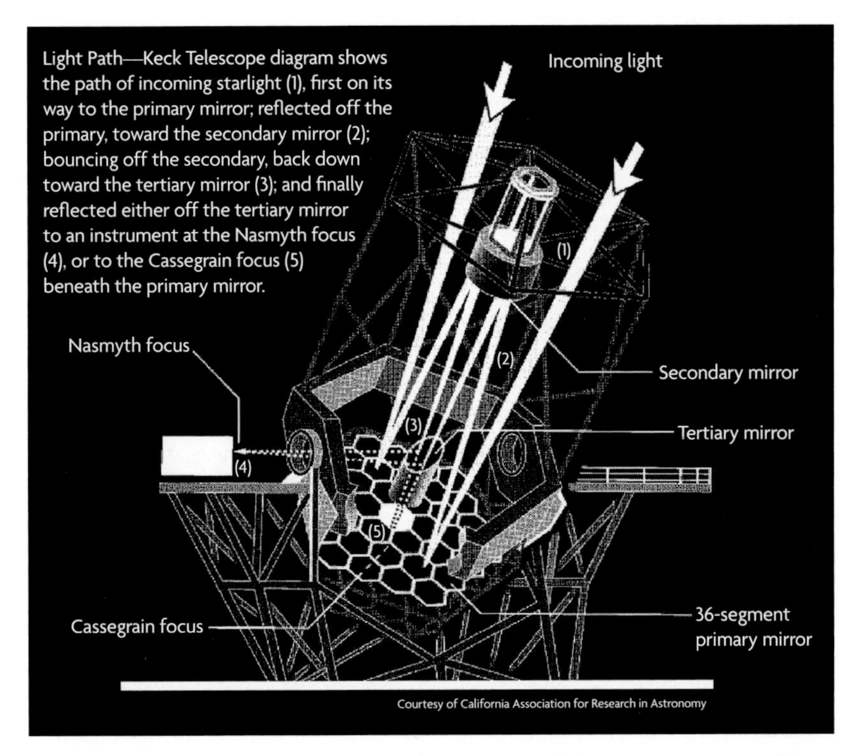

Figure 4.1 The 10-m Keck telescopes are characteristic of the new generation of large ground-based telescopes. One of them is illustrated here. Since the weight of the entire system is a result of the weight of the primary mirror, the primary mirror is composed of thirty-six lightweight sections held in alignment with electronic actuators. The entire telescope is so compact that it fits inside a dome barely larger than the 1-m refractor built at the Yerkes Observatory a century before (California Association for Research in Astronomy).

a small fraction of a monolithic mirror with the same light-gathering power.

Another modern feature of the Keck telescope is its mounting. As Earth rotates, all celestial objects appear to move across the sky in rotation about a line that is the projection of Earth's North Pole, the same process that makes the Sun appear to rise and set. For over two hundred years astronomers have known how to build telescope mountings that would correct for this apparent motion and keep the image precisely centered.

However, scaling up those kinds of mountings to a 10-m telescope would be prohibitively expensive. Again, innovation came to the rescue in the form of something called the alt-az (altitude-azimuth) mounting. This mounting is simplicity itself. The greatest load-bearing axis is vertical and the other axis is horizontal. Rotation about the horizontal axis allows the user to point to any altitude above the horizon, and rotation about the vertical axis lets the user point the telescope in any direction (the azimuthal angle).

How Is the Light Used?

Forming an image with an optical telescope is simply the first step in making an observation. Before photography you would simply use an eyepiece to transfer the light to the retina of your eye. Now things are quite different. If you want to check and adjust where the telescope is pointed, you look at a monitor that shows the results coming out of an electronic camera located where the eyepiece once was. These cameras have sensitivities comparable to the human eye. They can also make time exposures, so that you can locate very faint objects. This observation takes place in the comfort and greater efficiency of a remote telescope control room. What happens next is where the science begins.

Most of our information about objects in the sky is obtained by two methods, imaging and spectroscopy. Imaging is easy enough to understand, but spectroscopy takes a little more explanation (see Figure 4.2).

Imaging is accomplished by placing a two-dimensional detector, like a photographic emulsion or its modern equivalent, near the image formed by the telescope and simply letting the incident photons work their magic on the detector. It often isn't that simple because astronomers are usually willing to sacrifice some light in order to isolate the light containing the information they want. This isolation is achieved by using filters. In the case of stellar observation you can obtain a wealth of basic information by imaging them with three different filters that isolate sections of yellow, blue, and near-ultraviolet light. The filters take a large bite out of all the available light so that faint objects are made visible, even with a short exposure time. For an object like the Orion Nebula, where most of the light is concentrated into very narrow intervals of the spectrum, filters can isolate just the colors in the Nebula, rejecting all of the surrounding extraneous light.

Understanding spectroscopy requires an appreciation of how light is formed. Any object with a finite temperature emits energy in the form of electromagnetic radiation. Light is electromagnetic radiation, but only that portion that the eye can detect and those parts close to it. It is convenient in daily life to use names of colors to designate specific types of light, but the most quantitative way of measuring light is to use its wavelength, the distance between the crests of the electromagnetic waves. In units of one millionth of a millimeter (a length called the nanometer [nm]), deep red is about 650 nm; yellow, 580 nm; green, 500 nm; and the deepest violet, 420 nm. The wavelengths of radiation just beyond the deepest red are called infrared and those shorter than the deepest violet

Figure 4.2 This spectrum of the Sun illustrates the wealth of information contained in a high-resolution spectrum. The colors show how sunlight is actually composed of all the colors that the eye can see. The dark narrow lines are absorption lines caused by atoms in the coolest, outermost layers of the Sun. From a detailed analysis of these lines one can determine the temperature, density, and even the chemical composition of the Sun without actually going there.

are called ultraviolet. Our human eyes see light of wavelengths from about 420 nm through 670 nm.

Light has a dual personality. It is fully described as a continuous wave of radiation, but it also has the property of being a packet of energy called a photon. This means that light carries energy: the shorter the wavelength, the more energy carried by each photon. Infrared photons have less energy than photons in the visible spectrum and ultraviolet photons have more. When I refer to a photon having a given color, I don't mean that it is a discrete object that would have that color when viewed under a microscope, rather, that this is the color that would be sensed by the human eye.

About one hundred years ago the German physicist Max Planck described the properties of electromagnetic radiation in terms of wavelengths and photons. He demonstrated that a solid body or a dense gas radiates a continuum of photons and that the distribution of them depended critically upon the temperature of the source. A hot source emits most of its radiation at short wavelengths and a cool source at long wavelengths. The equation he used to describe this is now called Planck's law.

If all radiation came as a continuum of photons of different energies, the astronomer would probably be content to use filters for imaging, but this is not the case. The outer part of our Sun consists of a low-density gas. It is an atmosphere, like our atmosphere on Earth, but much hotter. All stars have atmospheres that modify the radiation coming from inside the star, leaving spectral signatures called absorption lines in the light we observe. These absorption lines contain valuable information about the properties of the stars' atmospheres, allowing us to determine the temperature, density, and even chemical composition of the star. We'll see in Chapter 7 that there are also ways of producing emission lines—the selective emission of radiation at precisely one wavelength—and this can be used to determine the properties of a low-density gas such as that found in the Orion Nebula.

There is a plethora of information available about the night sky if you have the tools to examine light in detail. Light can be divided into its component colors (a spectrum) by a device called a spectrograph. Even more information can be gleaned if light is divided into smaller bits called resolution elements. Higher resolution means greater spectral purity (the ability to measure light at smaller intervals of wavelength), but it also means that light from a given star is spread out and more difficult to measure. Because of this trade-off, many telescopes have an array of spectrographs, some of low spectral resolution, but able to measure faint objects, and others of high spectral resolution, able to determine great detail about brighter objects.

A typical spectrograph is usually able to sample only a small portion of an image, so that spectrographs don't use most of the information in the image formed by the telescope. However, spectrographs are employed the majority of the time on big telescopes because they produce the most helpful information about celestial bodies.

Detectors

Whether the astronomer directly records the entire image or samples a portion with a spectrograph, there is always a detector used to record the signal. The properties of the detector are just as important as the telescope in terms of effectiveness. In fact, a superior detector can surpass the performance of a larger telescope at a fraction of the development cost. This often happens, but never for long, because the proprietors of the bigger telescope usually quickly adopt the new detectors. As described in the preceding chapter, the advent of photography produced a fundamental change in our ability to make observations. However, photographic emulsions were never really efficient and it was difficult to obtain highly accurate results, especially when there was a wide range of stellar brightness.

Astronomers whose work emphasized photometric accuracy, the ability to measure brightness with great accuracy and over a wide range, investigated using a very different type of detector early in the twentieth century. This type of detector drew on knowledge of the newly discovered properties of the photon and the atom, expressed as something called the photoelectric effect (first explained by Albert Einstein). Some types of material will eject electrons when they are struck by photons. This occurs with a much higher quantum efficiency (ratio of recorded events to the number of incoming photons) on these surfaces than on photographic emulsions. If you can accurately record the number of electrons being knocked off the surface, then you have a direct method of measuring the number of photons that hit the detector. This can be done more accurately and over a wider brightness range than with photography. In the study of certain problems, like the light curve of variable stars, astronomers are not interested in the other stars in the image, but in studying one object with great photometric accuracy. In this process, the light-sensitive surface was packaged within a vacuum tube. In the early days the weak trickle of liberated photons were difficult to measure. However, during World War II the photomultiplier tube was developed. The photomultiplier tube stacks a series of electron-emitting surfaces within one vacuum tube, arranged so that the photon-induced electrons coming off the first surface strike a second similar surface, which produces a burst of electrons. This burst is then directed to a third surface, and so on. In the end each original liberated electron becomes a burst of a million or more electrons, which produces an easily measured current. After the development of the photomultiplier tube, photoelectric photometry became practical and has only been slightly improved upon in the last fifty years.

Even though the photomultiplier has wonderful sensitivity, it can only address a single object at a time. Astronomers are often interested in comparing objects in the night sky, however, which led to the invention of a generation of detectors that sought to combine the best of two worlds, that is, the high quantum efficiency of the photoemissive surface with the large size of the photographic emulsion. These detectors, called electron cameras, have photoemissive surfaces at their front, after which electrons are accelerated within a powerful electrical field. The high-energy electrons then strike a photographic emulsion, darkening it just like light, except with much higher quantum efficiency. The entire instrument had to be packaged within a vacuum, and the emulsions carefully retrieved while protecting the photoemissive surface from air (which would destroy it). These electron cameras were little used because they were so difficult to operate.

In parallel with these developments was an effort largely supported by the defense agencies to produce devices called image tubes. These also involved photoemissive surfaces in a vacuum, followed by acceleration of the electron bursts; however, in this case the high-energy electrons struck a phosphor screen at the back of the tube. The net result was that the faint image at the focal point of the telescope was amplified into a bright image on the phosphor screen. This bright image could then be examined if you were a soldier operating at night in Southeast Asia or reimaged onto a photographic emulsion if you were an astronomer. These image intensifier tubes were quite successful, but still limited by the low dynamic range of the photographic emulsion. Although vastly more rugged than the electron cameras, they too were a hybrid and doomed to a short career with telescopes.

Charge Coupled Devices (CCDs) are the most popular modern detector. These are solid-state detectors, being arrays of tiny, doped, silicon pieces called pixels (see Figure 4.3). As a photon enters the doped silicon it creates a charge deficit in that local area. During a time exposure these

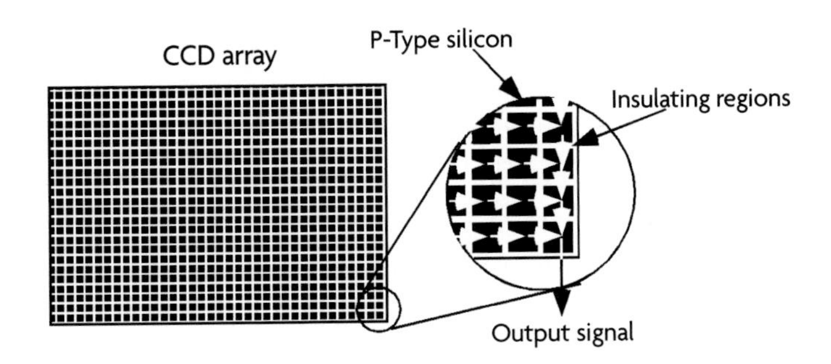

Figure 4.3 The most common form of modern detector is the Charge Coupled Device (CCD). A silicon chip is divided into an array of individual detectors (pixels). Light striking each pixel alters the charge in that area. At the end of the exposure the charge in each pixel is read out in bucket-brigade fashion, converting an image into a stream of electronic information, which can then be reconverted into an image on a display such as a television screen or a computer monitor.

electronic holes build up in number in direct proportion to the number of photons coming in and with quantum efficiencies that now approach unity (almost one charge hole created for every incoming photon). All this is wonderfully simple—the trick is in the readout. The potential image is actually a series of charge holes in a two-dimensional array of tiny detectors. The design involves dividing each pixel into three small sections, within each of which the electric voltage can be controlled. By sequentially changing these voltages, the signal from one pixel is transferred to its neighbor, whose signal has just been transferred to its neighbor, and so on, in bucket brigade fashion. Only low voltages and an accurate controlling clock are required to convert the electric image in the CCD into an electronic record that can be displayed as an image. The challenge lies in making the tiny detectors and their voltage controls, and making arrays of them large enough to be useful. These CCD detectors are ubiquitous and I expect that most readers of this book have operated them. No? Do you mean that you have never operated a home video camera or a digital camera? CCDs are the detectors in these devices. The CCDs that we use on large telescopes differ from the mass-marketed kind employed in video and digital cameras only in their having much lower background signals and being much larger in the size of the array. Of course this also means that CCDs used on telescopes are much more expensive, especially because the market for them is vastly smaller. In their use at the telescope they are commonly chilled with liquid nitrogen, which is an inexpensive, readily available refrigerant. At the liquid nitrogen temperature of −196 C (−321 F), the dark current (charge holes being produced in the absence of incoming photons) is reduced to almost zero. At the high temperatures in which home cameras are operated, the dark current is thousands of times higher, but it is small compared with the signals produced by the bright sources one records. CCD detectors have become the standard of our era. The size of available arrays has continued to increase, the largest now being about 4,096×4,096 pixels, which is small when compared with the potential of a full telescopic image. However, their great quantum efficiency and ease of operation have allowed them to displace photography in most applications, except for when a very wide-angle image is required.

Tuning In on the Universe

As mentioned earlier, light is only one small portion of a wide range of electromagnetic radiation. It is a very important portion because it is one of the two parts that penetrate Earth's atmosphere, thus reaching the surface of Earth, and the only portion our eyes detect. There is a second window in Earth's atmosphere, and it allows very long wavelengths of radiation to reach us. These are radio waves, which are as short as a few millimeters or up to more than ten meters in length.

The detection of such long-wavelength and low-energy photons presents particular challenges. The first detection of extra–solar system radio signals came in 1931. Karl Jansky of the Bell Telephone Labs serendipitously found that the center of our galaxy is a strong radio source. This was followed by a more complete mapping of the northern sky by Grote Reber with a telescope mounted in his backyard in suburban Chicago in 1936. Radio astronomy really came into its own as a tool for astronomers after World War II, when the technology for building sensitive radio detectors could find peaceful applications. It was discovered that many objects emit radio radiation by processes quite different from Planck-like radiation from the surface of a star. We now know that a low-density gas emits thermal radiation produced by the electrons and protons bumping into one another, a process we see occurring in the Orion Nebula, whereas other high-energy gases act like a particle accelerator, producing emissions like a machine called the synchrotron. Radio astronomy, one of the most successful and exciting subfields in modern astronomy, now produces a rich array of results, in spite of having a very important fundamental limitation: relative lack of image clarity.

The resolution (the image's angular size) of a telescope is determined by the ratio of the wavelength of the radiation being observed divided by the size of the aperture of the telescope. This fundamental property is established by the fact that there is an edge on the aperture and radiation is bent (diffracted) when passing an edge, which is called the diffraction limit. In this case smaller angular size means a better image. As I will discuss in Chapter 10, the shimmering of Earth's atmosphere typically allows the formation of images of about 1 arcsec angular extent (the full

Moon's diameter is about 1,800 arcsec). This means that an optical telescope of 10-cm (4 inch) aperture produces an image as sharp as that allowed by Earth's atmosphere. This size telescope is typical of that owned by a beginning amateur astronomer. However, if the radio astronomer is observing at a wavelength of 21 cm (a very frequent event since that is the fundamental wavelength of radio emission by hydrogen, the most abundant element in the universe), then he would need a telescope 43 km in diameter!

Figure 4.4 The Very Large Array (VLA) in New Mexico illustrates how radio telescopes can overcome their inherent low resolution by spacing components of the array at large distances. In this photograph all of the components are at their closest position. For maximum resolution they are spread out over 36 km (National Radio Astronomy Observatory/Associated Universities Incorporated/NSF).

To achieve the full image quality of its aperture, the reflecting surface (the aluminum mirror in an optical telescope, the wire mesh in the case of a radio telescope) must be smooth and of the right shape within a small fraction of the wavelength. This means that it is much easier to build a 10-m radio telescope than a 10-m optical telescope, but the angular size of the image would typically be several times the angular size of the Moon. The current state of structural engineering does not permit the building of enormous radio telescopes that would rival the quality of optical telescopes. However, if you use two or more telescopes in combination, the angular size of the image depends on the ratio of the wavelength of the radiation divided by the distance between these telescopes. This simultaneous use of two or more telescopes is called interferometry. The images are not as bright as if a single reflector had been used and it is difficult to extract an image from the signal that is measured; however, it can be done. The most powerful array of radio telescopes is the Very Large Array (VLA) located on the high plains of northern New Mexico. It is an array of twenty-seven individual radio telescopes (each 25-m in diameter) acting in concert and can produce images ten times better than optical images from ground-based telescopes (see Figure 4.4). There are even larger arrays of radio telescopes, capable of even better angular resolution, but the resulting images are extremely restricted. Unaffected by sunlight and moonlight, day or night, and at many wavelengths not affected by clouds, these telescopes can be used around the clock.

Astronomers without Calluses

Some astronomers don't go to mountaintops to observe or even use their computer terminals to submit observing programs to space instruments, yet they are valuable members of the field. These are the theoreticians and the modelers. There is not a clean division among theoreticians, modelers, and observers. In overly simple terms, one can look at the theoretician as a person using the most fundamental physics to intellectually explore what is happening in the universe. Usually, these researchers have a strong background in theoretical physics, often with no experience in the physics laboratory and none with a telescope. A good example of theoretical astrophysics is the prediction by the (then) physicist Fritz Zwicky, who predicted neutron stars only one year after confirmation of the existence of neutrons. Theoreticians have the great liberty of being able to make predictions about what might exist. It is usually thought that the most powerful support for a theory or hypothesis is when confirming evidence is found after it was predicted. Of course theoreticians also have a safe haven: they might say that you haven't found what they have predicted because you haven't observed in the right place or in the right fashion.

Modelers are the close cousins of theoreticians. They draw heavily on a strong background in physics and mathematics, but draw more heavily on observational results for inspiration and for exemplars of the objects they try to explain. Although most would consider Subramanyan Chandrasekar (Chandra) to be one of the best theoretical physicists of the twentieth century, the work for which he was awarded the Nobel Prize in Physics came from his model for white dwarf stars. The peculiar properties of these stars had been known for decades, but Chandra successfully applied new results in theoretical and experimental physics to this problem and demonstrated the correct model. He not only explained the objects that had been observed, but also made predictions. In particular, he predicted that star cores of more than 1.4 times the mass of our Sun would collapse to extremely small sizes. This complemented Zwicky's prediction of neutron stars. Chandra, therefore, represents an astronomer whose work fell both into the category of the theoretician and the modeler.

A better example of a modeler is Lyman Spitzer, who applied his in-depth knowledge of physics to two quite different subjects, the structure of dense clusters of stars and the interstellar medium of dust and gas. In each case he drew upon a large body of observational material and also new applications of basic physics to explain the structure and evolution of the globular clusters of stars and the gas between the stars. He did not try to apply known physics laws to radically new situations, rather, he drew the facts and the physics together.

Of course observers (those of us with callused hands from operating telescopes and building equipment) also engage in interpretive science.

Indeed, this is often the most fun, especially when you have made the observations yourself. Nevertheless, no one can cram into one lifetime the time needed to specialize in all areas of a field. The important thing is that there is a positive working relationship among the different types. In biology, the term "symbiosis" describes two quite dissimilar organisms living together to the advantage of both. That is the goal in the science of astronomy.

CHAPTER FIVE

Opaque Skies on the Clearest Nights

WHEN you are on a mountaintop or in the desert on a moonless night the night sky is almost overwhelming in its beauty. It feels like you can reach out and touch the stars and the sky is unimaginably black. With the spread of urban lights and increased air pollution such views are rare, and are the reason why big observatories are in remote locations. However, even the most ideal terrestrial locations are limited by the fact that all the electromagnetic radiation observed has to travel through Earth's atmosphere. This limits what we can see.

The Barrier to High-Energy Radiation

Our atmosphere is primarily composed of nitrogen and oxygen molecules with important traces of more complex molecules built of various combinations of carbon, hydrogen, nitrogen, and oxygen. High in Earth's atmosphere oxygen atoms change their grouping from the pairing found at the lowest levels (O_2, the "diatomic" form of oxygen) to a molecule composed of three oxygen atoms, called ozone. Ozone is poisonous in high concentrations, but isn't a direct threat to life because it is so high up. However, this ozone layer does affect life on Earth because it absorbs ultraviolet radiation of wavelengths between about 200 and 300 nm. It is the ultraviolet end of the spectrum of light that carries more energy. Exposure to this radiation without sunscreen can mean a nasty sunburn in the short term and possibly skin cancer in the long term. Sunscreen lotion looks clear once applied because it allows light of longer wavelengths to pass through, but blocks out the highest-energy ultraviolet light. The ozone layer is a quasi-permanent layer of sunscreen protecting the entire Earth. It blocks out the high-energy part of our Sun's radiation that would otherwise strongly affect survival of organisms on Earth's surface. The importance of the ozone layer is now widely accepted and has resulted in international agreements to prevent release into the atmosphere of chemicals such as fluorohydrocarbons in old refrigeration systems that can threaten the continued existence of this layer. Other, more stable molecules, such as diatomic oxygen and nitrogen, absorb radiation at even higher energies, so that if you're standing on Earth's surface, you cannot see radiation short of 300 nm.

Barriers to Low-Energy Radiation

Many molecules in our atmosphere are able to absorb the low-energy infrared radiation that arrives on Earth. The most common are H_2O and CO_2. H_2O essentially cuts off most radiation of wavelengths longer than 1,100 nm and continues to be important all of the way out to about 1 mm wavelengths. This barrier to infrared radiation is not as complete as the ultraviolet barrier, since the ability to absorb radiation jumps about with wavelength. Usually when H_2O isn't absorbing everything, another molecule like CO_2 does, but there are enough chance gaps that astronomers can observe infrared radiation at a few "infrared windows," even from the ground.

The very low-energy photons that are radio signals pass right through Earth's atmosphere. At very long wavelengths (above about 10 m), however, a new barrier becomes important—Earth's ionosphere, which is even higher (farther out) than the ozone layer. The ionosphere is a tenuous region of the upper atmosphere that has been modified by the Sun's radiation so that many of the molecules and atoms are charged because some of their electrons have been stripped off. The remaining atoms and molecules are called ions (hence, ionosphere). The ionosphere both absorbs and reflects long-wavelength radio radiation. The reflection property is what allows you to hear AM radio stations that are far away at night. The ionosphere's reflectivity actually increases at night so that the very long-wavelength AM radio signal that slants upward can be reflected and comes back to a spot over the horizon from the transmitter. In contrast, FM radio waves have a wavelength of about 3 m and that radiation simply goes out through the ionosphere and into space, with the result that FM radio can only be heard if you have a direct view of the transmitter.

The Atmospheric Windows

The three barriers described above bound two atmospheric windows in Earth's atmosphere. These are the visual (or optical) and radio windows. The visual window extends from 300 nm to 1,100 nm, whereas the radio window extends from about 1 mm to just beyond 10 m wavelength. How close you can operate to the edge of the windows depends on local conditions. For example, you usually cannot work down to the 300 nm visual window from observatories near sea level because scattering of light from all the molecules in the atmosphere becomes increasingly important and can shut the window within a few hundred nanometers of the ozone cutoff. Likewise, the short-wavelength end of the radio window is caused by H_2O, which means that only the driest sites can work at millimeter wavelengths.

It has been only natural that evolution has produced eyes on almost all advanced species. This exploits the copious amount of energy that comes from the Sun to give us information at a distance about our surroundings. The visual window is the one used by all seeing organisms because the photons arriving through the radio window carry very little energy and are quite difficult to detect. Moreover, in the preceding chapter we saw that the resolution of a telescope (or an eye) depends on the ratio of the wavelength of the radiation detected to the size of the aperture. This means that it would only be practical to have eyes sensitive to visual radiation, otherwise our eyes would have to be thousands of times larger than the rest of us. Since we evolved from sightless creatures, evolution took the easy path and we use light coming through the visual window. Even though the visual window extends from 300 to 1,100 nm, the human eye uses only the portion from 420 to 670 nm, the range over which most solar radiation is emitted. This means that the eye is an optimized combination of the properties of transmission of our atmosphere and the emission properties of our Sun.

This raises the tantalizing question of how evolution proceeds on other planets. Most stars are much cooler than our Sun. Planck's law implies that most of their emission will come out as lower-energy infrared photons, but these are hard to detect. Alien eyes would probably be optimized for the shortest wavelengths (highest-energy photons) that make it through their planet's atmosphere. If the atmosphere contains H_2O, this means the eyes could be near-infrared detectors and would probably be much larger than terrestrial eyes. However, it may be that in this circumstance evolution will take a very different path and use another method of detection of distant objects. The sonar system used by bats for operating in the complete darkness of caves is an excellent example of an alternative method of detection. Of course if there are such creatures out there they wouldn't be looking back at us because sound is not transmitted through the near-perfect vacuum of space.

The existence of two atmospheric windows allows earthbound astronomers to sample two regions of the electromagnetic spectrum. However, astronomical objects emit their radiation with complete disregard for the properties of our atmosphere. Planck's law defines the behavior of radiation emitted by a high-density gas or a solid body. There are special high- and low-energy processes that allow copious amounts of radiation at very short and very long wavelengths. This means that when we look at distant objects, we may be seeing but a small fraction of all the radiation that is emitted. For this reason astronomers have spent much of their energies of the last several decades in extending the range of the two wavelength windows and opening up new ones in the infrared and at high energies.

Nibbling Our Way into the Infrared and the Ultraviolet

Astronomers have only recently been able to exploit the entire visual window allowed by Earth's atmosphere. The natural sensitivity of photographic emulsions allows a good response to ultraviolet radiation, which the eye does not see, but it has been rare that we observe down to the 300 nm limit imposed by ozone. CCD detectors require special processing in order to be sensitive to ultraviolet, but such processing is now routine. It was much more difficult to observe in the infrared. Special photographic emulsions preprocessed in a careful way have come within a factor of ten of the blue sensitivity, but this still leaves photography at a disadvantage. The near-infrared is, however, the natural sensitivity regime for CCD detectors, which has allowed the full exploitation of the window right to the 1,100 nm limit.

We can push farther into the infrared region ordinarily blocked by atmospheric H_2O, even from or near Earth's surface, but we must go to an extremely dry site. Perhaps surprisingly, heading to a desert isn't the best way to escape water vapor because deserts often support large amounts of high-altitude water in its cloudless gaseous form. The driest observatory on Earth is at the Amundsen-Scott station at the South Pole. This is an incredibly dry location because of the constant low temperatures and the lack of circulation of the air into more temperate regions. Obviously, it is a difficult and expensive place to operate since the site is not only remote but high (2,847 m), and has an average temperature of about −49 C (−56 F) in the middle of the night.

An alternative way to find very dry conditions is through the use of high-altitude aircraft. Earth's lowest layer is called the troposphere, within which there is a free circulation of moisture-laden surface air. The top of the troposphere (the tropopause) occurs at about 50,000 ft at Earth's equator and slowly drops as one moves toward the poles. At mid-latitudes like the United States, the tropopause occurs at about 41,000 ft, which is an altitude that can be reached by airplanes capable of carrying telescopes. Such programs began in the 1960s with the flight of a small telescope located in the emergency escape window of a Learjet business aircraft. Next a 0.9-m aperture telescope was placed in a military C-141 aircraft. Named the Kuiper Airborne Observatory, this system operated for over twenty years, with several hours of pristine conditions during each of its flights. This observatory is being replaced with a 2.5-m telescope mounted in a Boeing 747 aircraft. Called SOFIA, this joint venture between the United States and Germany will be flying soon.

A Better View from Far Above

Space is really the best place for observing the universe and astronomy is beginning to exploit the capability of placing observatories there. However, observation from space is much more expensive than observation from ground locations. To look for information contained only behind the barriers outside of the visual and radio windows, we must be beyond the atmosphere. During the last half-century we have seen our capabilities extend from brief glimpses to quasi-permanent observatories and the wavelength range extended from the highest-energy photons (gamma rays) to the longest radio waves.

The first glimpses of the future came from a military rocket. During World War II the German government sponsored the creation of the first fully operational rocket, the V-2, a program led by Werner von Braun. A gifted engineer and leader, he had always been fascinated by space flight. His government's interest was in the creation of a suborbital rocket that could drop explosives on targets a few hundred miles away, but doing so meant traveling outside Earth's atmosphere during the peak of the trajectory. Hundreds of V-2 rockets were assembled from parts captured by the Allied forces at the end of the war and brought to the United States for study and application. Based at the White Sands, New Mexico government facility, these rockets were made available to scientists, engineers, and defense workers for studying the upper reaches of the atmosphere and the Sun, the only source bright enough for the detectors of the day. Fired almost vertically, the V-2 provided a few minutes of the view from space. By 1952, though the rockets had been expended and this program ceased, the potential of observing from space had been made plain. Experience with the V-2 also provided guidance for U.S.-developed small rockets, which were created as successors to the V-2 program. The most successful were used well into the period of orbital satellites. American plans for involvement in space accelerated when the Soviet Union successfully launched *Sputnik,* the first artificial satellite, in October 1957, shocking the American public. The realization that the Soviets had a technological capability Americans did not possess led to a surge of interest in building rockets that would travel into space. In particular, the government wished to create artificial satellites and carriers for delivering nuclear weapons to targets. By January 1958 the first U.S. satellite had been launched (on a rocket developed by the von Braun team, who had been working in America since the end of the war). Astronomers had been involved with the U.S. V-2 program and plans were immediately drawn up for building astronomical satellites as part of the fledgling civilian space agency, the National Aeronautics and Space Administration (NASA).

The first of these astronomical observatories was the *Orbiting Astronomical Observatories* (*OAO*). I was fortunate to have worked as an assistant in the earliest days of the OAO program. Each of these was to make observations in the ultraviolet region blocked by Earth's atmosphere. Three spacecraft were originally planned, with a progression of increasingly complex instruments, starting with simple stellar photometry and ending with high spectral resolution spectroscopy. The particular challenges of the *OAO* were providing stable pointing and electrical power to the instruments, and control from the ground. These considerations led to the development of a standard cylindrical spacecraft within which the different scientific payloads could be placed. The first observatory suffered a catastrophic power failure before any astronomical observations could be made, but a redesigned reflight was operationally successful (see Figure 5.1). The second observatory was lost through a launch error. The last of the series carried a high-resolution spectrograph and was highly successful. Scientifically it led to the design and launch (1978) of the *International Ultraviolet Explorer* (*IUE*), a joint enterprise with the European Space Agency (ESA). That satellite was placed in a geosynchronous orbit over the mid-Atlantic, where it could be alternately operated from Europe or the United States. The *IUE* was in almost continuous operation for nearly nineteen years and can be said to be the project that first brought space astronomy to a wide range of observers.

Figure 5.1 The first astronomical observatory in space (*OAO-A2*) is shown here just before launch. The octagonal cylinder is hollow and houses the hardware for operating the spacecraft, the telescope, and its auxiliary instruments. This was the first of three similar spacecraft employing progressively more sophisticated telescopes and instrumentation (Robert C. Bless and NASA).

Manned space flight had an astronomical role from the beginning. The highly successful X-15 rocket-powered aircraft traveled as high as 67 miles, well above the atmosphere, which allowed observing times comparable to the rocket probes. The X-15 program ran from 1959 through 1968 and in its last several years included an astronomy payload on a stabilized platform located aft of the cockpit. The *Gemini* program was the first to have the long duration and open access to space necessary

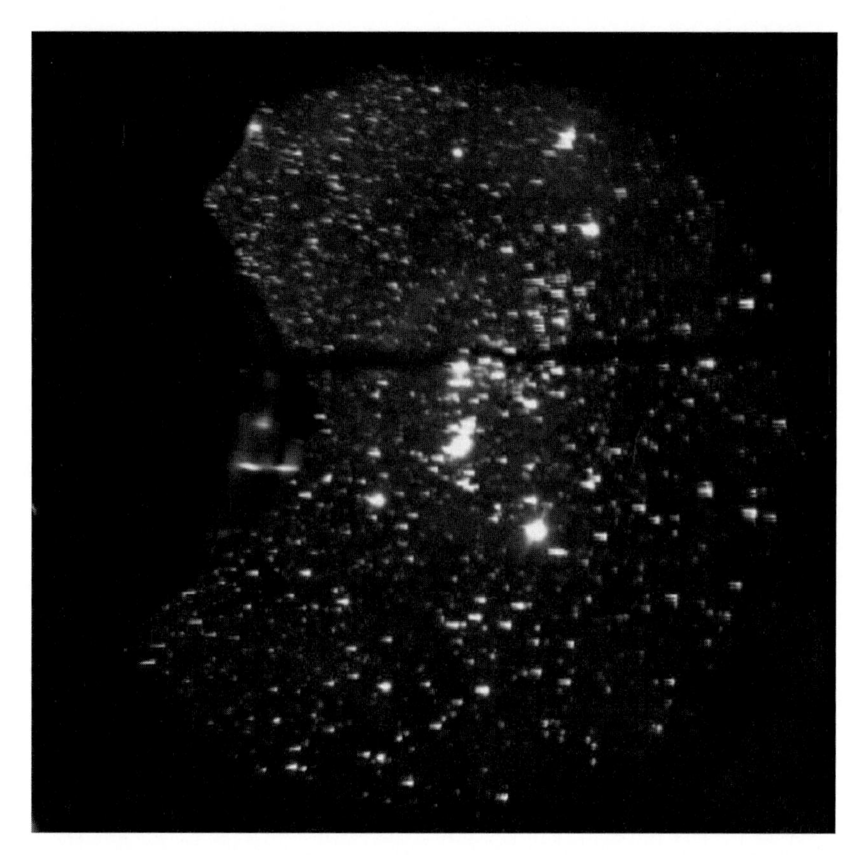

Figure 5.2 This is one of the first astronomical photographs obtained from space. The two-person *Gemini 11* spacecraft had just docked with the Agena rocket upper stage, which is seen in the left foreground with stars reflecting off of it. One of the astronauts opened his hatch and stood "up" to make this picture with film mostly sensitive to blue and ultraviolet light. The camera was pointed at the constellation Orion. The constellation is a bit hard to recognize because the antenna of the Agena partially obscures the middle and west members of the three belt stars. Because the film was more sensitive to blue light than the eye, the red star Betelgeuse appears unusually faint at the top middle, whereas the hot blue star Rigel stands out in the lower-right center. The hot young stars in the sword region appear bright in the middle of the image (Karl G. Henize and NASA).

for astronomical observations. The main purpose of that program was to develop spacecraft rendezvous capability and establish that man could function in space long enough for trips to the Moon; however, the program included "space walks." These "walks" were really excursions into the vacuum of space, made after opening the entrance hatches of the manned module. *Gemini 11* (1966) flew a photographic camera equipped with a thin prism that could disperse the light of stars into low spectral resolution spectra. This project was led by Karl G. Henize of Northwestern University, who later became an astronaut himself and subsequently perished during a climbing expedition to Mt. Everest. The *Gemini 11* expedition benefited many astronomers, including me: I was the lead author of the research article from the *Gemini 11* observations. The most successful of these observations was a two-minute exposure of the constellation Orion, which showed new aspects of the Barnard Loop nebula that surrounds the outer parts of the middle of the constellation (see Figure 5.2). Manned involvement with astronomical observations continues today. In Chapter 10, I'll describe how the Hubble Space Telescope, for which I served as project scientist for eleven years, could not have operated without the participation of astronauts.

Infrared astronomy also was a natural goal for space science. The first efforts were again made with sounding rockets, followed by dedicated satellites. The *Infrared Astronomical Satellite* was launched in January 1983 and mapped the thermal emission from cool bodies in the sky. Some infrared missions have been very limited (but profound) in their scope, such as the *Cosmic Background Explorer*, which mapped the sky at very long infrared wavelengths—wavelengths that helped astronomers trace the background radiation from the big bang phase of the evolution of the universe. The most ambitious infrared satellite to date has been the *European Infrared Space Observatory*, which weighed 2.5 tons upon launch in November 1995 and provided a wealth of photometric and spectroscopic data. All of the infrared spacecraft launched thus far have had relatively short life spans: the low temperatures necessary for their operation are achieved through the use of liquid or solid cryogens, which typically evaporate in about one year. In parallel with these many NASA and ESA projects there were many programs conducted with support from the

U.S. Department of Defense. Although obtaining astronomical information was not the goal of these missions, it was a by-product of the process. Most important to the research astronomer, these programs established new levels of performance for space telescopes and this technology eventually became available for use in unclassified programs.

CHAPTER SIX

Why Is a Star a Star?

A STAR could be most simply described as a gravitationally bound self-luminous body, although that definition wouldn't cover stars in some of their most extreme states. Stars are the result of a combination of basic physical processes that also occur here on the surface of Earth. We can study these processes in their local application and then see how they apply throughout a star's life span. As we'll see, stars are not as eternal as some romantic crooners like to sing about, but continuously change until reaching their final states.

Some Big Numbers

Naturally we know the most about our Sun, the closest star, although as a middle-aged star of relatively low mass, it does not shed much light on the birth or death of stars. Although its vital statistics are pretty mundane when compared with other stars, they are spectacular on a human scale. The luminosity (intrinsic brightness) of the Sun is 4×10^{19} (40 billion billion) megawatts. Putting this into perspective, the total human consumption of power on Earth is about 2 million megawatts. That 100-watt bulb in your lamp uses 1 percent of 1 percent of a megawatt. The mass of the Sun is 2×10^{30} kg, which is about the same as 1 billion billion billion large sport utility vehicles. Its diameter is 1.4 million km, which is 300 times the distance between New York City and Los Angeles. Rather than repeat these incredibly large numbers, in this book I'll continue the common practice of expressing the characteristics of other stars in the same scientific units used to measure the Sun, unless there is a good reason to do otherwise.

An aside on units is necessary at this point. Americans are among the last holdouts for the awkward British system of units. Almost all of the rest of the world uses the metric system. When we designate astronomical distances, in which the distances are enormous when expressed in Earth- or Sun-type units, the usual practice is to use light years, which are 63,241 times the average distance between Earth and the Sun, which in turn is 150 million km. The common system of temperature designation, the Celsius scale (C), is very convenient, but is still only relative (being tied to the properties of water, such as boiling or freezing points). The most meaningful temperature scale is one called absolute temperature (K for Lord Kelvin, who introduced it). At its zero point, objects have no heat whatsoever. This occurs at −273 C. It has the same increment size as the Celsius system, so that there is 100 K difference between the freezing and boiling temperatures of water in both systems.

The Sun is extremely hot. It is 5,780 K on the surface and in its middle 15.5 million K. The surface temperature of the Sun is duplicated only with great difficulty on Earth (never naturally), and the interior temperature of the Sun is matched only for a small fraction of a second during a nuclear reaction, such as the explosion of an atomic bomb.

Starting with Gravity

An understanding of what goes on within stars starts with the explanation of the motion of our Moon and the planets. As I've already mentioned, early in the sixteenth century Nicolaus Copernicus published a Sun-centered model for the solar system that employed circular orbits. This model was given support by the discovery of moons circling Jupiter by Galileo Galilei at the beginning of the seventeenth century. Through the use of precise observations of the motions of the planets made by Tycho Brahe, Johannes Kepler was able to develop a much more accurate model for our solar system. He established that the orbits were ellipses and could give a general expression for the fashion in which the orbital velocity varies in different parts of an orbit, concluding this work in 1619. All of Kepler's work was phenomenological, i.e., it established a model, but did not present any underlying causal principles.

The explanation for the motion of objects in our solar system was provided by Sir Isaac Newton in the late seventeenth century, following his successes in developing the laws of mechanics that describe all motion. Newton showed that there must be a hitherto unrecognized property possessed by all bodies, because they could attract one another even if not connected.

This attractive force was dependent upon the masses of the two bodies and varied inversely as the square of the distance between them. This was a most remarkable step, for Newton was positing a force that could act at a distance without anything connecting the bodies, a very revolutionary concept. We now call this Newton's Universal Law of Gravitation and it takes the form

$$F = G\, M_1\, M_2/r^2$$

where F is the attractive force, r is the distance between the two bodies of masses M_1 and M_2, and G is a normalizing constant that is dependent upon the units being employed. Gravitational force is very weak and only becomes obvious when bodies are large. We are held to the surface of Earth by the attraction of our bodies toward the center of Earth. Newton showed that the magnitude of force that could explain the rate at which bodies fell toward Earth could also explain why the Moon was held in orbit where it is.

Newton first established the principle that an object in a stable orbit was in equilibrium, with two balancing forces. Gravitational force wants to cause the Moon to fall toward Earth, but this attractive force is balanced by an equal but opposite outward force—centrifugal force. Figure 6.1 shows how this works. Centrifugal force arises from the fact that a moving body's momentum wants to carry it straight ahead, which means away from the other body. The Moon's straight-ahead momentum gives it an apparent outward force, which is balanced by the force of gravity.

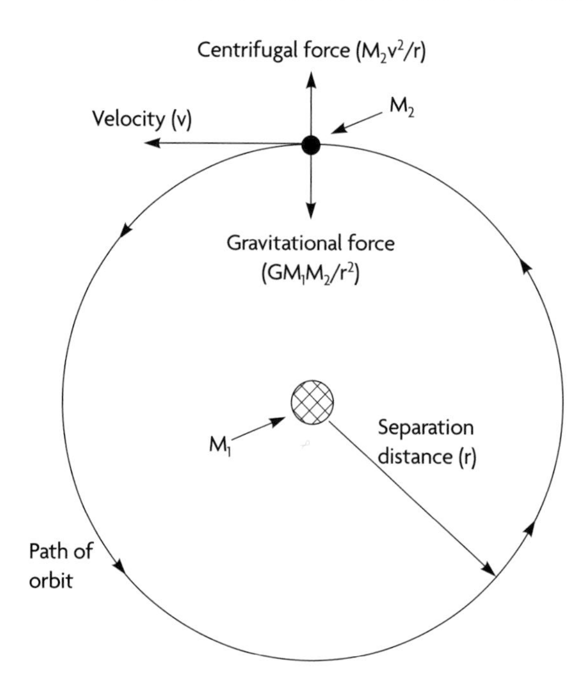

Figure 6.1 Objects can orbit about one another if the forces acting on them are in balance. In the case of large objects, the attractive force is from gravity, whose magnitude depends on the masses involved and the distance between the objects. The outward (centrifugal) force depends on the mass of the moving object, its velocity, and the distance from the central object. Once this equilibrium is established, the orbit is essentially stable forever, unless a third body disturbs the situation.

For objects with elliptical orbits, where they are more distant at the tips of the ellipse, the objects need only move slower (less momentum) to balance the weaker force of gravity there.

This one simple law (universal gravitation) in a stroke explained the motion of the planets about the Sun and the moons around their host planets, giving the underlying explanation for Kepler's empirical laws of motions of the planets. The Universal Law of Gravitation is just that—universal—which means that it operates on all bodies everywhere. In some cases it isn't the most important force. The electrons in orbit about the nucleus of an atom are held there through the electrostatic attraction of the two dissimilar charges of the electrons and the nucleus. The molecules in our body and in the solid material of Earth are also held together by electrostatic charges operating at the atomic level, with gravity playing no important role. However, our bodies as a whole are electrostatically neutral, so that we are bound to Earth by gravity.

Why the Atmosphere Stays Around

Except for environmental issues about the chemical composition of our atmosphere, we take it for granted. We don't worry about it escaping into space or being collapsed down to a scale where it hovers about our ankles, leaving us gasping. This is because Earth's atmosphere is in equilibrium, just as much as the Moon in orbit is in equilibrium. The inward force is again gravity. The Earth is attracting every molecule in the atmosphere, trying to compress it toward the center of Earth. In this case, the outward force is not centrifugal force caused by momentum but something quite different. An object in orbit is in mechanical equilibrium, whereas our atmosphere is in hydrostatic equilibrium (hydro deriving from a term for the motion of a fluid).

When there is a difference in pressure of a gas, the atoms or molecules in that gas move from where the pressure is high toward where it is low. A good example is when a tire is punctured—the air escapes. This really means that the gas is moving toward hydrostatic equilibrium. The air low in our atmosphere is of much higher pressure than the vacuum of space, so that there is a continuous outward force toward space. Fortunately, this outward force is balanced by the inward gravitational force of Earth. This condition is called hydrostatic equilibrium. The details of this process mean that the density (hence pressure) of the atmosphere is highest at the bottom of the atmosphere and slowly drops as altitude increases. This is why it is so hard to breathe at higher elevations and why jet airplanes fly faster at high altitudes.

Holding a Star Together

This same process of hydrostatic equilibrium is what holds stars together because stars are also gaseous. Unlike Earth's atmosphere, where almost all the atoms are bound up in molecules, the temperatures of stars are high enough that if there are molecules, they will only be found in the outermost parts. The other important difference from our atmosphere is that stars are gaseous throughout. The Earth's atmosphere is of negligible mass and is gravitationally trapped on the surface of a solid body. In contrast, stars have no portion that is solid, so that there is a continuous change from the outer atmosphere right down to the center of the star. A star is a sphere gravitationally bound to itself. At every point inside the star there must be enough gas pressure to hold up all of the material that lies above that point. Since the amount of pressure is the product of the gas temperature times the number of particles per unit volume, the star's local pressure is a result of both. The farther into the star you go, the more material there is lying above that point, so that the pressure continuously increases toward the center. This means that the temperature and density of material increases, so that in the case of our Sun, the central temperature increases 2,700-fold at the center and the density has changed from 0.02 percent that of the air we breathe to 148 times the density of water. This dramatic variation is all due to the gravitational force of the star on itself. If the total mass of a star is lower, then the central temperature will be lower and the density will be less. If the total mass of the star is higher, these parameters increase dramatically.

Sustaining a Star

Since stars are emitting light that carries away energy, there must be some method by which the star replaces this loss, because otherwise stars would have very short lifetimes. Our Sun would be able to sustain itself only a small fraction of the time that we know from the geologic record it has existed. The source of this energy lies in the nuclei of the atoms that form the interior of the stars. The effects of self-gravity are so great that they produce temperatures and densities high enough that the nuclei can actually collide. It takes extreme conditions for this to happen because the nucleus of an atom is always positively charged and two bodies of the same charge repel one another. However, at high enough temperatures each nucleus is moving fast enough to overcome this repulsion and actual collisions occur. In the case of hydrogen, which is the most abundant element in our Sun and most other stars, collisions of hydrogen atoms and its subsequent products can ultimately lead to the buildup of one helium nucleus out of four hydrogen nuclei. The important bottom line of this celestial alchemy is that the helium nucleus has 0.7 percent less mass than the hydrogen that made it. The missing mass has been converted into energy (E) via Einstein's famous equation, $E=mc^2$, where m now is the missing mass and c is the speed of light. Although the amount of energy released in one cycle of converting four hydrogen nuclei into one helium nucleus is small, there is so much material in the Sun that this process can sustain the Sun for twice as long as its present age.

The more massive the star, the more extreme the conditions will be in its center, and hence the faster the rate at which nuclear fuel–burning will occur. Figure 6.2 shows how rapidly a star's luminosity increases with its mass. The mass of a star is difficult to derive and can only be derived directly for stars paired off into binaries. What we can do for all stars is to determine their surface temperatures from their spectral types and to find their luminosities from measuring their apparent brightness and then finding their distances. When this was first understood, long before the process of nuclear fuel–burning was discovered, we saw that most stars fall along a single path in a diagram of surface temperature versus luminosity. This is illustrated in Figure 6.3, later in this chapter. We now know that this path is the location of stars of different mass as they are in their stable equilibrium state of converting hydrogen into helium. The first forms of this diagram were derived independently and at just about the same time by Ejnar Hertzsprung, a Danish astronomer, and Henry Norris Russell, an American astronomer, at the beginning of the twentieth century. This type of figure is now known as the Hertzsprung-Russell diagram, or H-R diagram. The location on the H-R diagram of stars while they are burning hydrogen in their cores is called the Main Sequence. Since this is the longest phase of a star's life, this is where most stars are found.

What Happens When the Fuel Runs Out?

The process of nuclear fuel–burning means that stars are continuously changing through a well-defined life cycle. This process is called stellar

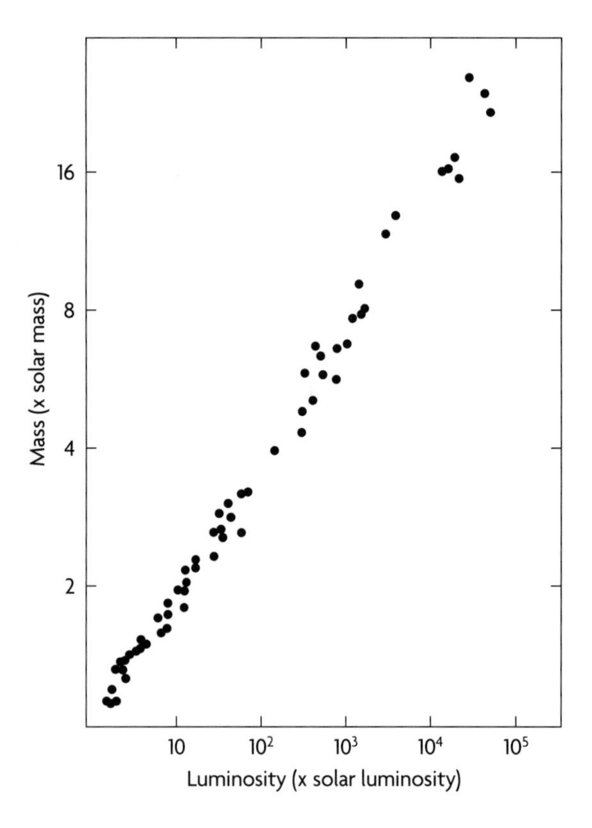

Figure 6.2 Stars that are more massive get hotter at their centers because of their greater gravitational force. This causes them to burn their hydrogen fuel at a much greater rate. The disproportionate consumption is illustrated in this diagram showing the mass and luminosity of stars on the Main Sequence. Note that a star of ten times the mass of our Sun is about 2,000 times as luminous.

evolution even though it describes the changes in an individual object, whereas biological evolution describes changes in a group of related specimens.

Stellar evolution proceeds very slowly at the beginning, with the star slowly building up a core of the helium that was formed from burning hydrogen. Once this helium core has formed the hydrogen-burning process shifts to the outer surface of the core where there is sufficient temperature and pressure to burn hydrogen. As hydrogen burning continues in the shell, the inside of the helium core shrinks, causing the center of the core to become hotter and denser.

Up to this point, all helium in the star has remained inert. The reason for this is that the helium nucleus has a double positive charge (hydrogen has a single positive charge). This double charge means that it is more difficult to have a successful collision of helium nuclei because the repulsive forces are greater. However, when the temperature becomes about 100 million K, the helium in the center of the core begins to burn, forming carbon (which has the same composition as three helium nuclei) as a by-product, and the star alters its structure. Now it is burning two fuels—helium in its core and hydrogen in a shell at the surface of that core. Eventually a carbon core builds up and helium burning occurs in a nearby shell, with hydrogen burning farther out in a cooler shell. Stars of our Sun's mass or of a few times greater will then run out of nuclear fuel because they won't become hot enough to burn the by-products of helium burning. But if a star is more massive, the central temperatures will be large enough that the heavy elements like carbon are burned in a succession of steps. High-mass stars are the most exotic, since they end up looking like an onion, with multiple layers of different composition and fuel-burning processes. These high-mass stars can then form a core of iron. Whatever the original mass of the star, it will eventually run out of available fuel. In the lower-mass stars this is because they never get hot enough to burn the products of the previous stage of fuel burning and in the high-mass stars, fuel burning stops because burning iron actually removes energy. In any case, the star eventually collapses. The end product of the highest-mass stars is a black hole, of a high-mass star, a neutron star, and of stars like the Sun or a few times more massive, a white dwarf.

The time scale over which all this happens is very different for stars of different masses. The Sun has enough mass that it will last about 10 billion years. A star five times more massive (a 5 solar mass star) will have five times more fuel. However, it will have a luminosity 380 times as great (hence will be burning fuel at 380 times the rate of our Sun). It will run out of fuel in a time 5/380 that of our Sun, i.e., 130 million years. This pattern continues, with a 10 solar mass star lasting for 40 million years and a 37 solar mass star, like the brightest star in the Orion Nebula Cluster, having enough fuel to last only 3 million years, a time comparable to the evolution of man!

How Old Are Stars?

As the interior of a star becomes more complex as it evolves, it moves across the H-R diagram, as illustrated in Figure 6.3. This movement for massive stars is like a progression of two steps forward, one step back.

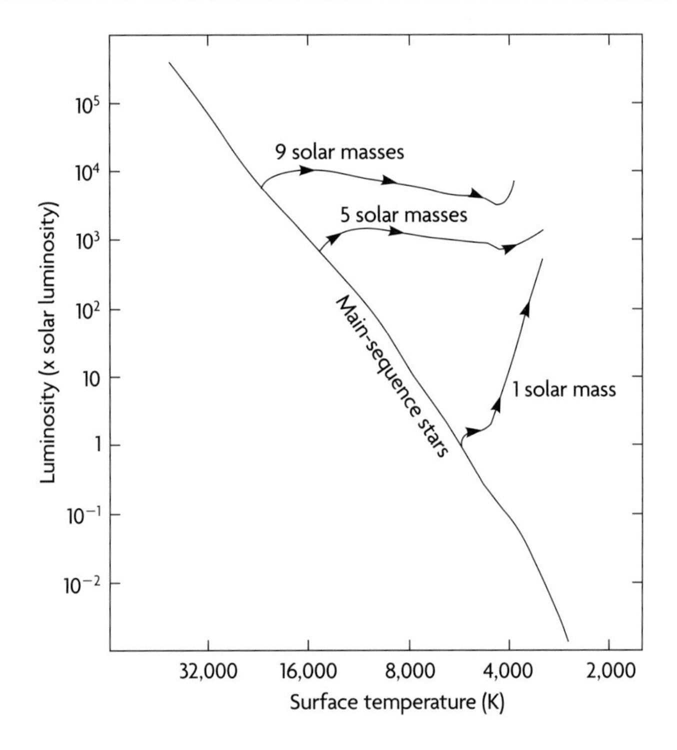

Figure 6.3 Stars adjust their size and luminosity and move from their original positions on the Main Sequence when they begin to deplete the fuel supply of hydrogen in their central cores. Since the more massive stars are intrinsically more luminous, they use up their fuel first.

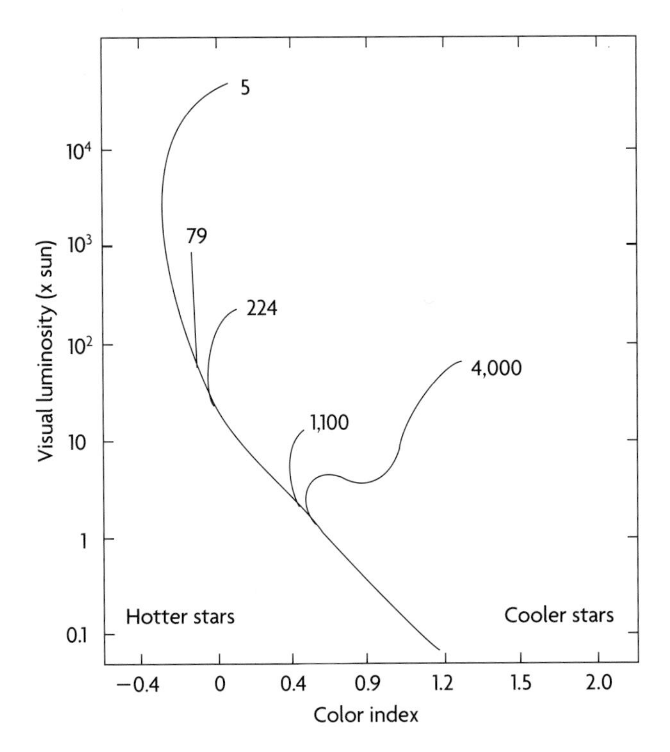

Figure 6.4 The color-luminosity diagram is a variation of the H-R diagram (increasing brightness is toward the top and increasing temperature toward the left). The Color index is a measure of the blueness (hot) or redness (cool) of a star. The different lines indicate the locations of stars in different clusters. Stars within a cluster have the same age, but different masses. Since the more massive stars evolve off the Main Sequence faster, the distribution of the stars varies with the cluster's age. These ages are indicated for each cluster in units of millions of years.

The star moves across the diagram at about the same luminosity while becoming cooler on the surface. Each of the small steps back correspond to the ignition of yet another new source of nuclear fuel. The length of time between each of the steps is shorter and shorter. For a star about the mass of our Sun, the motion is much more complex, even though there are fewer dance steps. When the star is in its hydrogen shell–burning phase it cools only slightly, but expands enough that it increases 10,000-fold in luminosity until ignition of helium in the core begins. At that point the star dramatically collapses back to a position near the Main Sequence, then retraces a very similar track to its first ascent while undergoing simultaneous helium and hydrogen burning, although the time for this second ascent is much less than for the first.

Fortunately, there is a way of determining the ages of stars if they are in a cluster. All of the stars in a cluster are nearly the same age. In the early hydrogen-burning phase stars don't jump around in the H-R diagram, rather, they smoothly move away from the Main Sequence. Clusters are formed of stars of a variety of masses and the rate at which a star moves off the Main Sequence is faster the more massive the star. This means that stars in a cluster of a given age will be spread out in a way that indicates their age. Stars having Main Sequence burning times for core hydrogen that are shorter than the cluster's age will have evolved off the Main Sequence, while the stars with longer Main Sequence burning times for core hydrogen will all still be sitting on the Main Sequence. The stars that have evolved the most will be farthest from their original positions.

This means that if you look at the H-R diagram for a cluster of stars, you can easily determine its age. You simply determine the point at which the least massive stars have moved off the Main Sequence. The results for a host of clusters of stars in our galaxy are shown in Figure 6.4. The lines for each cluster stop near the top because stars even more massive than those stars have evolved fast enough that they are quickly moving or have reached their final states, which are always very low luminosity. In Figure 6.4 the oldest cluster has an age of about 7 billion years, whereas the youngest is only about 5 million years old. The oldest clusters of stars in our galaxy are the rich globular clusters, where one finds stars about 10 billion years old.

In summary, astronomers now have a good understanding of how stars operate and evolve. This knowledge has allowed us to determine that different stars have different ages, and that star formation must be a continuing process.

C H A P T E R   S E V E N

Bengt Stromgren's Spheres

WHEN one looks at the Milky Way on a dark night it is obvious that it is not smooth; rather, it has irregular dark bands along its middle. Whether or not these dark areas were devoid of stars or whether there was something obscuring stars was debated for more than a century. At the beginning of the nineteenth century, Sir William Herschel proposed that there was material spread out between the stars that was blocking light from more distant stars, but he could not prove that this was the case. It was only in the 1930s that it was firmly established that the region between the stars is not a perfect vacuum, but is filled with gas and dust. We are not talking about great quantities of material. The average gas density is about one atom per cubic centimeter (cm^3), whereas Earth's atmosphere has a density of 2.5×10^{19} molecules per cm^3. The absolute amount of dust would not worry the most fastidious housekeeper, for there is on average only one dust particle per million cubic meters. However, the distance between the stars is enormous and the total amount of dust and gas is large, being several percent of the mass of stars in the Milky Way Galaxy. The dust manifests itself by blocking and reddening the light from very distant stars (much like fog obscures our view of distant objects) while the gas forms signatures in the spectra of distant stars.

It is when the dust and gas are concentrated at higher densities that things really become interesting. Soon after the development of the astronomical telescope it became obvious that there were some diffuse clouds of light that didn't break down into a multiplicity of stars even with the most powerful instruments. Although many thought that these were indeed clouds of material in space, this couldn't be proven. The skeptic could always argue that this cloud was really a host of very faint stars, each too faint to register individually. The strongest indirect argument for their being clouds was that of Herschel, who found that one class of nebulae ("clouds" in Latin) always had a single star at the middle. He argued this demonstrated that the diffuse material could not be stars. The situation became clear in the middle of the nineteenth century when Orion and then other nebulae were examined with spectroscopes and were shown to have emission lines, just like excited gases in a labora-

Figure 7.1 T. E. Houck photographed the Southern Milky Way from South Africa in the 1950s with a special "all-sky" camera designed a decade before. This shows how the band of the Milky Way is marked by numerous clouds of obscuring material, the most striking being the small "Coalsack" feature just to the right of the topmost leg of the three legs holding the camera.

Figure 7.2 Edward Emerson Barnard was a major figure in observational astronomy in the late nineteenth and early twentieth centuries. In spite of having had almost no formal education, he is credited with many astronomical discoveries. He used a photographic technique to determine the ubiquity of interstellar clouds of dust and gas (Yerkes Observatory, University of Chicago).

tory. Some of the nebulae did not have such spectra and it was only when photographic emulsions had become sufficiently sensitive that it was seen that the nebulae without emission lines had a continuous spectrum remarkably similar to nearby stars.

Barnard, Hubble, and the Galactic Nebulae

By World War I there was an enormous compilation of photographic images of the nebulae, which was largely the result of the efforts of a most remarkable person, Edward Emerson Barnard (see Figure 7.2). Born in Nashville, Tennessee in 1857 into conditions of dire poverty, Barnard had only a few months of formal education and went to work as a photographer's assistant at the age of nine. Bright and incredibly hardworking, he soon became enamoured with astronomy and acquired a progression of small telescopes. He benefited from contacts with the astronomer at the local, newly founded Vanderbilt University. He first made his mark by discovering a comet, then numerous others. Such discoveries were considered remarkable in those days and there was prize money given for confirmed finds. Slowly Barnard worked his way into the milieu of contemporary astronomy and landed a position at Vanderbilt, which he left soon after (1887) to become a staff member of the newly established Lick Observatory in California, which boasted the world's largest telescope, with a 36-inch aperture refractor. Possessing acute vision and powers of observation, he gained worldwide attention by discovering the fifth moon of Jupiter in 1892. He was able to draw on his background of photography to begin systematically recording the appearance of the band of the Milky Way using progressively more powerful wide-field cameras. By 1895 he had transferred to the Yerkes Observatory of the University of Chicago and completed his photographic compilation of images of not only the star fields of the Milky Way, but also the many nebulae that dotted it.

While at Yerkes, Barnard came to know a graduate student, Edwin P. Hubble, who entered the graduate program in 1914. Hubble too was not out of the ordinary mold, but in a very different way. He had been a Rhodes scholar after his undergraduate years at the University of Chicago and had taught high school for a year before coming to Yerkes to pursue what really interested him. Upon completing his thesis on nebulae (Barnard was on his examining committee), he deferred accepting a prestigious position at the Mt. Wilson Observatory. Instead, he went with the American Expeditionary Forces under General Pershing to Europe. When he did arrive at Mt. Wilson, where the 100-inch Hooker reflector had just gone into operation, he continued the work he had begun with his dissertation. In 1922 he published his first paper, which made him famous. He demonstrated that the nebulae lying along the Milky Way were basically of two types. The first was the reflection nebulae, in which the spectrum was essentially identical with that of the illuminating stars. The second was the emission-line nebulae, in which the spectrum was dominated by narrow emission lines, with only a weak underlying continuous spectrum. Moreover, he demonstrated that the difference between the two types of nebulae depended upon the temperature of the dominant nearby star (see Figure 7.3a and b). If the star was hotter than about 25,000 K, then the nebula would be of the emission-line type, and if the star were cooler, then the nebula would be of the reflection type. This was a purely phenomenological division, without an explanation of the causal physics, but the work provided the basis for clarification later.

Two Types of Galactic Nebulae

Reflection nebulae mimic the radiation from a nearby star. They do this by simply scattering the light from that star. The scattering is caused by minute dust particles, characteristically about 0.1 micron in size (one micron is one thousandth of a millimeter). Particles this small are very effective in scattering light because they have a very large surface area for each unit of mass. We see the effects of light scattering in many ways. The glow of the sky just after sunset is from sunlight being scattered by molecules and dust particles in our atmosphere. In this case there is not enough material to block out the view behind it, so that we can see bright stars and the inner planets in this glow. When the material is very thick, like in a thunderstorm cloud, we only see a bright edge around the cloud, which again is caused by scattered light. When we look in our Milky Way we see the same type of phenomena. Some reflection nebulae are thin, so the glow is diffuse, whereas in others we see only the bright rim around the cloud.

The light from emission nebulae is dominated by emission lines from atoms. The process that produces this radiation is not the scattering of light, but a process called fluorescence. You might be familiar with this process: a party light that makes white clothes seem to glow in the dark is a good example of fluorescence. In that case a special light bulb is emitting most of its light as ultraviolet radiation. Although the photons in that radiation are high energy, our eyes don't detect them. However, when those high-energy photons strike white cloth, the ultraviolet photons are absorbed and through a series of events produce visual-wavelength light, so that those shirts and blouses seem to glow in the dark for no obvious reason.

A similar process is happening in the emission-line nebulae. In this case it is primarily hydrogen atoms that are performing the fluorescence in two steps. To understand these steps one has to understand the basics of how an atom is built.

Figure 7.3a NGC 6611 is an excellent example of an emission-line galactic nebula. In this case there are many stars as hot as Theta1C Orionis, so that a much larger volume than the Orion Nebula is photoionized. The Orion Nebula would set into the central bright region of this object, which is more than four times as distant as Orion (T. A. Rector and B. A. Wolpa, National Optical Astronomy Observatory/Association of Universities for Research in Astronomy/NSF).

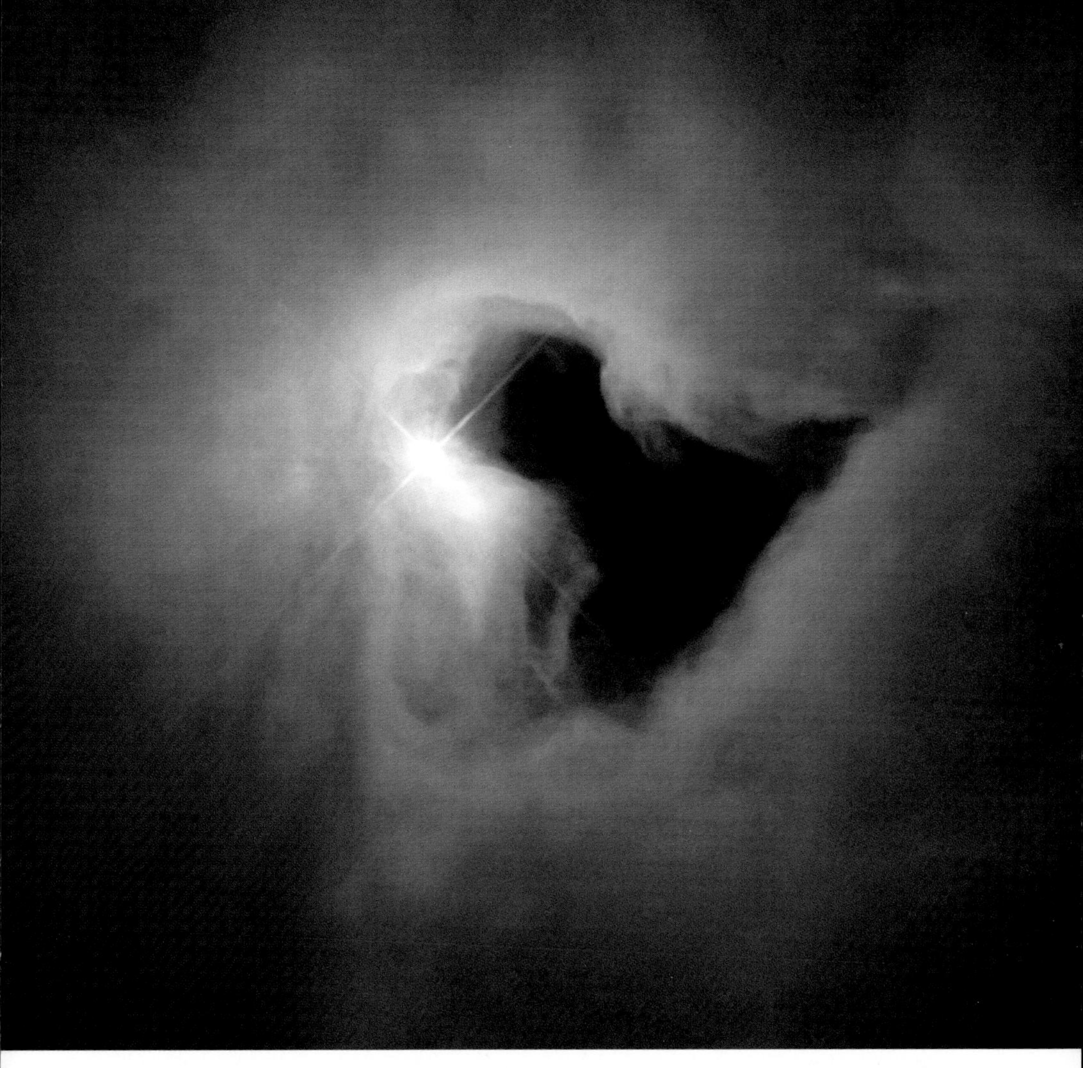

Figure 7.3b NGC 1999 is very different. Its dominant star is relatively cool so that fluorescence does not occur; rather, the nebula is visible because the dust mixed with the gas scatters the starlight. The dark, thick T shape is a dense cloud of molecular gas and dust lying in front of the nebula. Its longest dimension is only 0.04 ly (the author, the Space Telescope Science Institute, and NASA).

The Structure of an Atom

Atoms are composed of a massive central body called a nucleus and one or more electrons in orbit around that nucleus. In a sense they are similar to planets in orbit around our Sun. There are, however, several important differences. In the case of the solar system the attractive force is gravity, whereas in an atom the attractive force is electrostatic (arising from the fact that electrons have a negative charge and a nucleus a positive charge). The other important difference is that atoms are so small (about one ten-thousandths of a micron) that they are affected by quantum mechanics. One result of quantum mechanics is that particles like an electron aren't only simple little spheres, but also have the property of waves, much as electromagnetic radiation has the characteristics of both waves and particles (photons). A by-product of this property is that only certain discrete orbits of the electron are allowed. This is in marked contrast to planets in our solar system, which can exist in a continuous distribution of distances from the Sun.

Let's consider the simplest atom, hydrogen, which is depicted schematically in Figure 7.4. The electron can exist in any of the possible or-

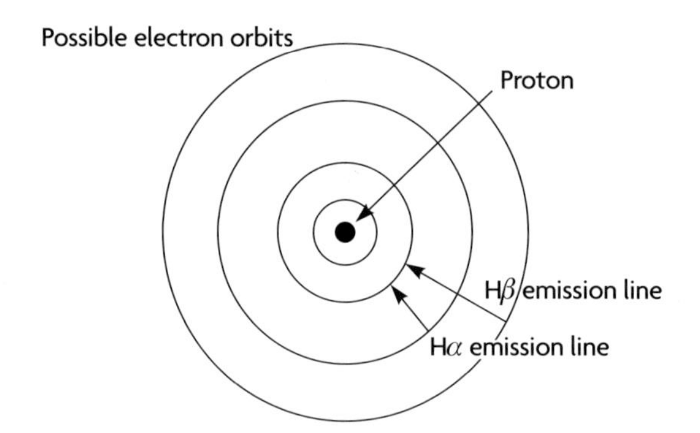

Figure 7.4 Quantum mechanics determines that the electron in orbit about the proton in the center of a hydrogen atom can only occupy discrete orbits. This means that when ultraviolet fluoresces, the energy is concentrated into a few emission lines, like those labeled here. This is a highly efficient mechanism for converting invisible ultraviolet light into visible light.

bits. The ones closest to the nucleus will bind the electron the most and those farther out are more weakly bound. This means that an electron moving from an outer (weaker binding) orbit to an inner (stronger binding) orbit will give up energy by making the change. When this loss of energy occurs, a photon is emitted. Since the difference in the two orbits is very exact, this means that an emission line is formed. If an electron moves from an inner to an outer orbit, on the other hand, energy must be given to the electron, which means that a photon of that exact energy must be absorbed. The difference in energy between the second and third orbits corresponds to red light, in particular, the so-called Hα line. In general, electrons want to be in close orbit, which means that typically the hydrogen electron will be found in its $n=1$ orbit.

There is a second way of moving an electron to a higher orbit and this is if the atom collides with a free electron. In that case the electron is moved out during the collision, but immediately after drops back to a lower orbit.

Other atoms are more complex than hydrogen. Whereas the nucleus of hydrogen is composed of a single (positively charged) proton, other atoms can have additional protons and also massive neutral particles called neutrons. Often but not always the number of neutrons is equal to the number of protons. For example, the helium nucleus is composed of two protons and two neutrons. The number of electrons that can be held in orbit about the more complex nucleus will normally be the same as the number of protons. A neutral atom is produced when the nucleus is surrounded by the same number of electrons as there are protons in that nucleus. If you remove one or more electrons, then the resulting atom is called an ion. Ions have all the normal properties of an atom, except that their electron orbits are more tightly bound, since you have removed some of the like-charged electrons.

Making Ions by Absorbing Light

The outermost orbits of an atom have a very discrete energy above the innermost orbit. In the case of hydrogen, this energy is 13.6 electron volts. An electron volt (eV) is the amount of energy that an electron acquires if it moves between two plates having a difference of voltage of one volt.

An eV is a very small quantity of energy. A visual light photon has an energy of about two eV. Electrons with energies above that of the most extreme orbit will no longer be bound to the nucleus and will freely move about, and the atom will have become an ion. The limiting energy of the outer orbits is called the ionization energy.

The most common way to remove an electron from an atom is through absorbing a photon with energy greater than the ionization energy. This means that if a hydrogen atom absorbs a photon of radiation that has more than 13.6 eV, then the hydrogen atom will have become a hydrogen ion. The process of removing electrons by absorbing photons is called photoionization, a term I'll use often.

Changing Ions Back into Atoms

As the ion and electron freely move about, they sometimes come close enough together that the negatively charged electron is recaptured by the positively charged ion. It almost certainly won't be the original electron from that atom, nor does it have to come from the same type of atom. Any electron will do. Since electrons seek their lowest orbit, this means that if the electron combines to a higher orbit, that reconnection will be followed by the electron cascading down between the higher orbits. At

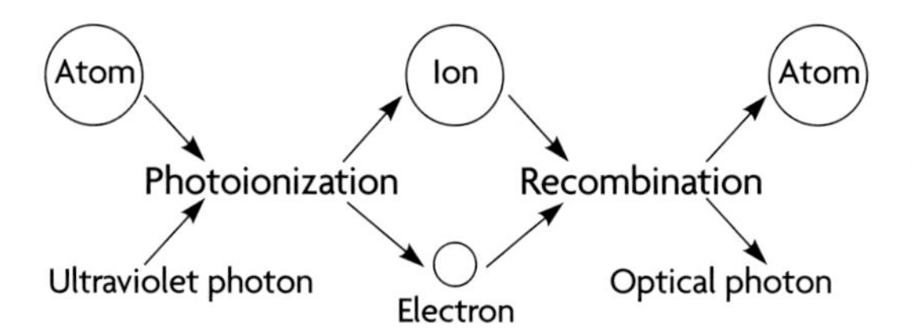

Figure 7.5 The process of fluorescence of ultraviolet light by an atom involves a number of steps. Stellar ultraviolet photons destroy (photoionize) the atom by ripping off the electron, leaving a positively charged atom called an ion. Later the ion and the electron can recombine to form the original atom and give off optical photons in the process.

each step it emits a new photon, for example, that red Hα photon. This process of combining an electron and an ion is called recombination.

For a gas in equilibrium the rate at which photoionizations occur will be exactly equal to the rate at which recombinations occur. Although this makes it sound like an empty exercise, it is not. The process of photoionization followed by recombination has converted a high-energy ultraviolet photon, which we cannot see, into many lower-energy photons, which we can see. The two steps together are called fluorescence (see Figure 7.5).

Explaining the Mystery of Nebulium

Things fell together nicely in the middle of the nineteenth century for spectroscopists. Spectra of the Sun revealed a wealth of absorption lines. We now know these to be caused by electrons moving from an inner orbit to an outer orbit. As various atoms were isolated and put into gaseous form, the same lines could be generated in the laboratory, thus demonstrating that the stars were composed of the same stuff as Earth. However, one set of related lines in the Sun defied identification. It was posited that the Sun had a special element not found on Earth. This element was designated helium (after helios, the Greek word for Sun). Later, helium was found as a trace gas here on Earth. We now know that helium is the second most abundant element in the universe and its underabundance on Earth is a result of its chemical inertness. The hydrogen we have retained is largely that which was bound up in water, whereas the helium remained unconnected and quickly evaporated from the early Earth.

A similar situation arose in the study of the nebulae. Even during the days of prephotographic observations of their spectra it was noted that the most intense lines were often a pair of green emission lines. The ratio of the strengths of the two lines was always the same. Sometimes they were about the same strength as the Hα line, but more often than not they were even stronger. One could easily explain the other strongest lines, because they all arose from hydrogen, but these lines and others found only in nebulae remained a mystery. As photography pressed the spectrographic process to detect fainter and fainter nebular lines, more lines of hydrogen were found but most of the new lines were unrelated to what one saw from the common elements in the laboratory. These were ascribed to an element unique to nebulae called nebulium. This interpretation became more and more uncomfortable because it meant that this mysterious element was a major component of the nebulae.

The mystery was solved in 1928 by a spectroscopist at the California Institute of Technology, Ira Sprague Bowen, whose major interest was in the spectroscopy of atoms and ions in the laboratory and the identification of the detailed orbits that produced them. What he found was that he could identify orbits in quite common atoms and ions that had the same spacing as the energy of the nebulium lines. For example, those intense green lines corresponded to differences in the orbits of oxygen from which two electrons had been removed. Beyond these coincidences he was able to provide the reason why the nebulium lines were not seen in the laboratory.

When an electron is sitting in a higher orbit, it will want to spontaneously drop down to a lower orbit. This usually occurs in an incredibly short time, typically less than a millionth of a second. However, the details of quantum mechanics demonstrate that transitions between a certain small fraction of the orbits will happen only very slowly, typically taking several seconds to happen. These transitions are called "forbidden" and the corresponding emission lines are called forbidden lines. To show their special nature, they are indicated within square brackets. For example, emission from doubly ionized oxygen is designated as [O III]. The Roman numeral after the symbol for the element is the notation of the ionization. Neutral (atomic) states are designated as I, singly ionized as II, and so on. Since a normal transition occurs more than a million times a second and a forbidden-line transition occurs only about every second, it means that the forbidden lines can be suppressed if there is an alternate way of moving between the orbits. Nature provides that alternative in collisions with free electrons (most of which come from the photoionization of hydrogen). If collisions with free electrons occur more rapidly than a few times per second, this means that the bound electrons will be knocked into a lower orbit and the corresponding energy will be given to the free electron rather than coming out as radiation. Bowen pointed out that in the nebulae the densities of electrons were so low that the forbidden-line transitions between orbits would be the principal way for decays to occur. In the laboratory the densities were so high that the free electrons knocked the bound electrons about in their orbits and no emission lines were generated. Bowen's research explained a mystery of three-quarters' century duration and showed that the nebulae were actually composed of the same material as our Earth, although not necessarily in the same relative abundance.

Bengt Stromgren Ties It All Together

Henry Draper could not have imagined that the Harvard College Observatory in the 1930s would be the center of the application of the newly developed science of quantum mechanics to objects like the Orion Nebula. In a series of papers, Donald Menzel and others developed the theory of how a gas would behave if illuminated by a nearby hot star. They fully understood that the dominant mechanism that determined the nature of the gas would be the process of photoionization, followed by recombination, and that the energy put into the gas through photoionization would be the source of the gas temperature. However, these gifted theoreticians, who dealt principally with the minutiae of physics, were not closely allied with observational astronomers, who were more likely to concentrate on the big picture.

Meanwhile, a more observational approach was being taken at the Yerkes Observatory, under the directorship of Otto Struve. He was the last of a family line of distinguished German/Russian astronomers. Struve rejuvenated scientific research at Yerkes, which had become a quiet place after its founding director, George Ellery Hale, left to establish the Mt. Wilson and (later) Palomar observatories in southern California. Not only did Struve begin to hire young men who would become major figures in mid-twentieth-century astronomy and astrophysics, he also entered into an alliance with the University of Texas for the construction of a major observatory in West Texas. Boasting an 82-inch telescope, then the second largest in the world, the McDonald Observatory was a much better place for observing than Yerkes's location in southern Wisconsin.

One of the first instruments Struve built at McDonald was a unique spectrograph, designed especially for work on emission-line nebulae. Struve recognized that if an emission line subtended a large angle, then the speed of the spectrograph would not depend on the size of the telescope that imaged the nebula onto the spectrograph's slit. The Yerkes astronomers and opticians came up with a unique instrument. It eliminated the need for a telescope and was a pure spectrograph pointed directly at the sky. It employed one of the first examples of a Schmidt camera as its means of recording the spectrum, formed by a prism. The advantage of the Schmidt camera was that it could image a wide field of view and yet do this with superb image quality and high speed. This "nebular" spectrograph was soon showing that emission-line nebulae were to be found all along the Milky Way. Some of these nebulae were like the first ones discovered (Orion being the very first) in that there was a core of very high surface brightness material. However, there were far more low surface brightness and much larger nebulae. They all shared the characteristic first pointed out by Hubble that they were associated with hot stars.

Bengt Stromgren was a young theoretical astronomer at the University of Chicago. He had spent the last parts of his first two years at Yerkes, where the nebular spectrograph was designed, constructed, and tested piggyback on the 40-inch refractor. He benefited from this association with the observers and with other young theoreticians such as Chandrasekhar and Gerard P. Kuiper. In 1939 he published "The Physical State of Interstellar Hydrogen" in the *Astrophysical Journal,* which was probably the single most influential paper in the study of emission-line nebulae. It led to a new term in astronomy, "Stromgren spheres." Stromgren developed a model for when a luminous hot star found itself within an extended cloud of neutral interstellar hydrogen. He demonstrated that the photons from the star that were capable of photoionizing hydrogen, i.e., the photons with an energy above 13.6 eV, would all eventually be absorbed and destroyed by causing a photoionization. The gas

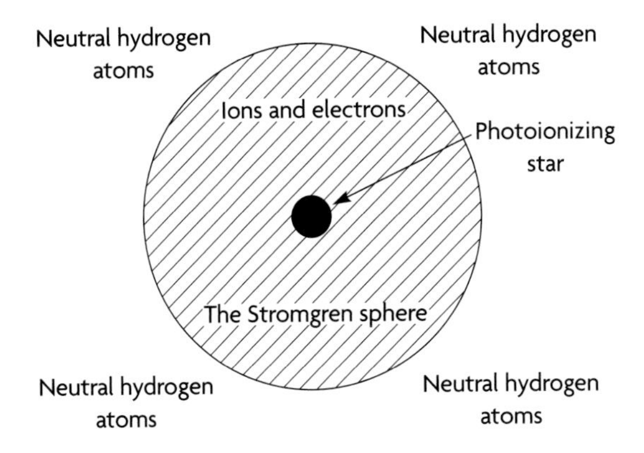

Figure 7.6 A Stromgren sphere is the zone of ions and electrons formed around a hot star embedded in a large cloud of neutral hydrogen atoms. This is what produces objects like NGC 6611 shown in Figure 7.3a.

near the star would be almost completely ionized, a state that would continue out to a specific distance, at which point the ionization would abruptly drop to being completely negligible. Since photoionization is always followed by recombination, which produces visual emission lines like Hα, the ionized zone would be a glowing sphere of radiation with a sharp edge. This is just what one frequently observed (see Figure 7.6). Stromgren also understood from Planck's law (which governs where radiation is emitted by a hot body) that the hotter the illuminating star, the greater the fraction of radiation that was above 13.6 eV. It is this high-energy radiation that determines the size of the nebula. A hotter, more luminous star would form a larger ionized zone (now called an H II region) than a cooler, less luminous star. In fact, if the star's temperature was lower than approximately 25,000 K, there would be so few ionizing photons coming from the star that the resulting H II region would be negligible in size. Likewise, if the density of the surrounding gas was higher, then the photons from the star would be absorbed more readily and the H II region would be smaller. In a single paper Stromgren could explain the diversity and similarities of the appearance of the emission-line nebulae and why they looked different from the reflection nebulae.

The latter are made of the same stuff as the emission-line nebulae; however, the illuminating star simply doesn't produce a detectable amount of photoionization.

Stromgren's paper was almost too successful. Its predictions were so comprehensive yet simple that astronomers applied the model to too many emission-line nebulae, including the Orion Nebula, thus slowing the recognition of the true physical nature of that object.

CHAPTER EIGHT

The Explorers Set Sail

THE Orion Nebula has been a beacon for astronomers since the advent of the astronomical telescope. Sherburne Burnham noted in 1889, "Perhaps no object in the sidereal heavens has received more attention from astronomers than the multiple star Theta Orionis [at the core of the Orion Nebula]. It has been the subject of careful study by the most eminent observers, provided with the best astronomical instruments; and the relative positions of the principal stars have been determined with the greatest possible accuracy." Burnham was writing about the discovery of the first of the objects we now know are very new stars with protoplanetary disks. His statement about the concentrated attention of astronomers to this one small section of the sky remains true today and it seems that Orion is always one of the first places studied when a new technology is developed or a more powerful instrument built. In this chapter I'll explain how our picture of the Orion Nebula has evolved during the fifty years of post–World War II observational astronomy, a period of unprecedented growth.

Early Results from Palomar Mountain

The 200-inch giant telescope at Palomar became operational in 1947–48. I can remember as a sixth grader reading about it in "Weekly Reader," a newspaper written for schoolchildren. In fact, it is likely that these articles are what produced the theme for an assigned essay, "What I want to be doing in 25 years." I recall saying, "I want to be an astronomer observing with the 200-inch telescope at Palomar Observatory." Fortunately, it didn't take twenty-five years for me to achieve that goal.

What I couldn't have known then was that the 200-inch was merely the last of a series of "biggest and best" telescopes in the world built by George Ellery Hale. Nor could I expect that in the future I would occupy one of the jobs that he had held, director of the Yerkes Observatory, and at about the same age (in our mid- to late twenties). There we might say the comparison ends, but, writing about Hale now, it is strange how lives intertwine. When Hale created the Yerkes Observatory in 1897 the telescope that got the most attention was their giant 40-inch refractor, which was then the world's largest telescope. There had been larger aperture telescopes, such as the speculum metal reflector built by the Earl of Rosse in Ireland in the 1840s. The problem with these giant telescopes was that the metal used for the mirror, an alloy of bronze, rapidly tarnished and required refinishing as often as every few months. Also, metal reflectors were very heavy. Hale built a prototype telescope at Yerkes in the 1890s, a 24-inch aperture reflecting telescope using the newer technology of a silver-on-glass mirror, and housed it in one of the smaller domes. Glass is much less dense than metal, so that the mirror and hence the entire telescope was much lighter. The thin silver reflecting coat also tarnished in a few months, but it was simple to recoat it with chemical processes. The telescope was a tremendous success because it packed an instrument with 60 percent of the aperture of the great refractor into a dome a small fraction of its size. Clearly a new path for construction of telescopes had been established. This led Hale to construct a 60-inch and then a 100-inch telescope at Mt. Wilson, in southern California. His greatest achievement was the funding and construction of the 200-inch at Palomar.

The design and construction of the 200-inch telescope extended over a period of twenty-one years, so that leadership for the project had passed to others as Hale aged (he died in 1938, age 67). The joint directorship of the Mt. Wilson and Palomar Observatories passed to Ira Bowen in 1945. Bowen, mentioned in Chapter 7, researched those nebulium lines in the spectra of gaseous nebulae. It was only natural that he would recognize the power of the 200-inch for spectroscopy. From the beginning he made sure that the Palomar telescope was equipped with the most powerful spectrograph of its day (see Figure 8.1). This instrument used an array of four giant gratings mounted in a large living room–sized thermally isolated chamber in the bowels of the observatory dome. The light for the image was piped to it through a series of flat reflecting mirrors. The spectrograph was a substantial fraction of the size of the telescope itself and attaching it directly to that instrument was simply not an option.

The combination of the world's best spectrograph and the world's larg-

Figure 8.1 Spectrographs are probably the most significant means of studying what is happening in an object like the Orion Nebula. It is not surprising that the famous spectroscopist, Ira Bowen, made sure that the world's most impressive telescope of its day, the 200-inch giant at Palomar Observatory, also had the most powerful spectrograph. This precompletion drawing by Russell W. Porter shows how the spectrograph was larger than most telescopes. Note the rather formally dressed astronomer at the upper right of the drawing (Palomar Observatory).

est telescope meant that the early results were superb. Because of the scientific interests of staff members Guido Munch, Donald E. Osterbrock, and Olin Wilson, much of the spectroscopy time was spent studying gaseous nebulae, and very naturally the Orion Nebula got its fair share of attention. One of the first studies, conducted by Wilson and Munch, was to determine the velocities of the material in Orion using the Doppler effect.

Christian Doppler was a nineteenth-century natural scientist who determined that the wavelength of waves appeared to change in a fashion depending on the relative motion of the observer and the emitting source. This is a universal relation and applies to both sound and light. When standing alongside a highway, cars approaching us have a higher pitch sound than after they have passed and are moving away. In the case of light, if the emitting material is moving away from us, the light is shifted to a longer and redder wavelength (thus red-shifted) and when the emitter is moving toward us, the light is blue-shifted. The percentage shift of the wavelength is the same as the percentage of the velocity of light of the relative motions (Doppler's Law). Since we can measure or calculate the intrinsic wavelength of nebular gas in the physics lab (therefore at rest with respect to the observer), then a comparison of the observed wavelength with the at-rest wavelength gives one the velocity of the nebula.

Wilson and Munch used the 200-inch spectrograph and Doppler's Law to map the velocities of many positions across the face of the Orion Nebula. They did this in multiple emission lines, including the ubiquitous hydrogen lines, but also many of the forbidden lines of various ions (lines that previously had been ascribed to nebulium). They mapped the velocities in samples that were about 1 percent of the total size of the nebula. This allowed Munch to determine that at the finest scales the gas in Orion was moving as if it were turbulent. Turbulent motion is random motion, but the packets of material within the gas behave in a related fashion. This is described by Kolmogorov's law, which relates the size of the packet to its random motion. Kolmogorov's law seems to work in a wide variety of conditions, from Earth's atmosphere to the Orion Nebula.

While Wilson and Munch were mapping and interpreting the fine-scale motion within Orion, Osterbrock was working with Michael J. Seaton, a British theoretician, to refine a technique that would allow astronomers to determine the density of nebulae emitting certain types of forbidden emission lines called doublets. These doublets are pairs of lines arising from two very close orbits in a heavy element (like oxygen, sulfur, chlorine, and carbon). The same processes that put electrons into these orbits also compete with radiation in depopulating those orbits. As a result, the intensity of the two lines directly changes with the density of the gas. Osterbrock and Seaton derived the theoretical relation between the doublet ratios and the density of the local gas. This meant that a simple observation of the relative strength of two nearby lines (in a doublet) could tell you how dense the material was, without ever needing to sample that material. This theoretical work was one of the most influential steps in the study of the physics of nebulae. Osterbrock then observed the oxygen forbidden doublet in the ultraviolet (373-nm wavelength) and determined the density of the material across the face of the nebula.

Astronomers can also determine the temperature of a nebular gas by looking at the ratios of certain other forbidden emission lines. In this case the lines come from widely separated orbits and their ratios are almost independent of the density, but very dependent upon the electron temperature. In the case of the Orion Nebula, the most common temperature is about 9,200 K. These two techniques, taken together, mean that the primary characteristics of a gas—its density and temperature—can be determined remotely. An application of the same approach also allows one to determine the abundance of the elements within the gas.

Epicycles on the Stromgren Model of Nebulae

When the wrong model for the solar system, the Earth-centered Ptolemaic system, was in use, the inherent simplicity of the model (the Sun, Moon, and planets going in circular orbits around Earth) was gradually eroded as observations of planetary positions became more accurate. This led to the introduction of secondary motions on top of the primary motions. These small circles imposed on the greater circles were called epicycles. These complications were joined later by other refinements necessary to keep the model consistent with observations. Such a pattern is usually a sign that the basic model is flawed, but it often happens that models are made progressively more complex until something happens that causes a paradigm shift to a new and simpler model that can explain all the observed features. In the study of our solar system this shift in paradigms came with the acceptance of the Copernican, Sun-centered model. We'll see in this section and the next that a similar pattern was followed in modeling the Orion Nebula. In the case of the study of nebulae, the analog of the Ptolemaic model was the Stromgren sphere. All nebulae were thought to have a constant density, and consist of spherical clouds of ionized material surrounded by neutral material.

Radio astronomy blossomed following World War II when those who built electronics used their skills for this peaceful application. The fundamental limit was the low resolution of these instruments (due to the large ratio of the radio wavelengths to the size of the telescope, as discussed in Chapter 4). Eventually these telescopes began to have just enough resolution that they could begin to see the Orion Nebula as a real object and not simply an unresolved point source of uncertain size. In the case of a gaseous source, the brightness in any one direction will be proportional to the square of the density of the gas times the thickness of the gas in that direction, a quantity known as the emission measure. The high-resolution radio observations were able to provide values of the emission measure across the face of the nebula.

Osterbrock combined the results from his mapping of the densities of the Orion Nebula with the radio telescope determinations of the emission measure. What he found was that the densities didn't behave anything like the assumptions of Stromgren's model. In particular, the density dropped steadily from a central value of about 10,000 atoms per cubic centimeter the farther away from the region of the Trapezium stars he looked. Since the emission measure depends on the thickness of the nebula and the density and Osterbrock now knew the density, he could derive the thickness along each line of sight through the nebula. For a Stromgren sphere this thickness would be greatest in the middle, where the brightness is greatest, then slowly drop until reaching almost the outer boundary of the sphere, where the thickness would drop abruptly to zero. Osterbrock quickly abandoned the assumption of a constant density and calculated the thickness of each line of sight, using the local values of the density. The answers didn't make sense. He then assumed that the nebula was spherical in form and that the density dropped progressively when going outward. The only way that he could reconcile the density and emission-measure determinations was to add an additional feature to the model. Drawing on the clumpy appearance of the nebula, he modeled the gas as grouped into small clumps that were spherically distributed. The density of the clumps dropped when moving outward in Osterbrock's model. The inherent simplicity of Stromgren spheres had failed miserably when applied to the most famous gaseous nebula. Simply retaining the assumption of spherical symmetry meant introducing new assumptions. It was like adding epicycles onto the Ptolemaic model of the solar system. It was logical, but incorrect.

Blisters Instead of Bubbles

Theoretical studies of the motions of the ionized material were conducted soon after measurements of the velocities within Orion and other gaseous nebulae became available. A Stromgren sphere should not be a permanent and stable object. This is because when the gas around a hot star is ionized and the hydrogen atoms are changed into free protons and electrons, the number of particles in the gas doubles. Likewise, ionization followed by recombination leaves some of the star's energy in the gas, heating it from the original temperature to about 10,000 K. As mentioned in Chapter 6, the pressure of a gas depends on both the density of particles and their temperature. Photoionizing and heating the gas means that the act of forming a Stromgren sphere produces a beautiful sphere of ionized gas but one with a pressure about 200 times higher than its surroundings, so that the Stromgren sphere begins to expand. The expected expansion velocity in a Stromgren sphere is about half the velocity of sound in that gas, which means the sphere should be expanding at about 8 km/s. This kind of expansion velocity (detected through measurement of the Doppler shift) does occur in several gaseous nebulae, in particular those whose forms indicate that they are closely approximated by Stromgren spheres. In other cases, including Orion, the motions simply don't look like those expected for a Stromgren sphere. In fact, it could be shown that if the spherically symmetric model for the nebula was correct, it should blow itself apart in only about 10,000 years, which means that either the model is wrong, the calculations are wrong, or that we were born at just the right time!

The Stromgren sphere model for gaseous nebulae assumed that the hot central star formed in a homogeneous medium of cool neutral material. By the 1970s it was understood that new stars are formed in molecular clouds, which have from 100 to 100,000 solar masses of material. One of the best-defined molecular clouds is an elongated affair running down the direction of the Orion Sword stars. Since these clouds contain both molecules and dust, we experience an observational selection effect. This means we will selectively see as optically bright objects those gaseous nebulae formed near the surface of the host molecular clouds and on the side that faces us. A gaseous nebula formed deep within a molecular cloud would be invisible to the optical observer because of the large amount of overlying foreground dust.

A new set of theoretical models was calculated. These models predicted the dynamical behavior of such nebulae formed near the surface of a molecular cloud. It was found that in this case the over-pressure situation of the Stromgren sphere would essentially blow off any overlying neutral material and the sphere would quickly empty itself into the surrounding low-density region at velocities several times the velocity of sound. This sudden but brief explosion is called the champagne phase. What is left behind at the end of the champagne phase is a thin ionized blister of material on the far side, i.e., up against the high-density host molecular cloud. Almost all of the ionization occurs in this thin layer on the surface of the molecular cloud, which is called the ionization front. The ionization front is where the transition from neutral to ionized gas occurs. It is analogous to a fire front moving across a dry open field of grass—the unburned part of the field is the molecular cloud, the ionization front is where the burning is occurring, and the burned-out region is where the original Stromgren sphere was located.

The ionization front will also be in an over-pressure situation. On one side is the high-density but cold molecular cloud and on the side facing the observer (outward) there will be little material. We would expect material to be accelerated as the ionization front passes through it, because of the pressure difference. The material right at the ionization front would then have the velocity of the molecular cloud, and the material farther out would have blue-shifted velocities, with the farthest material being blue-shifted the most. This velocity difference is exactly what we observe in the Orion Nebula, where the low-ionization ions close to the ionization front are indeed at the velocity of the host molecular cloud, whereas the velocities increase to about 10 km/s of blue shift at the ions closest to us.

Once the correct interpretation of the progression of velocities among different ions had been made, it was obvious that the Orion Nebula belonged to this post-champagne phase of nebulae. What we call the Orion Nebula is actually a thin blister of ionized gas on the surface of the Orion Molecular Cloud. Osterbrock's variation of density with distance from the middle reflected the fact that more gas is being photo-ionized nearest the central stars. The decrease in brightness away from the center was simply a result of those outer regions receiving less ultraviolet light. The shift in paradigm occurred and the complicated, spherically symmetric model was replaced by the blister model rapidly and almost completely.

Building a Three-Dimensional Model of the Orion Nebula

When we look at astronomical objects we see them from so far away, compared with their sizes, that they look flat. Some objects may look three-dimensional, but this is an illusion and it is just as likely that the side that appears closer is actually farther away. The technical way to say this is that the objects are too far away for us to see them as having parallax (where the objects are in subtly different directions, as observed with binocular vision). This fundamental property of observing distant objects means that when we develop a three-dimensional model of an object, it is just that, a model. However, in the case of a blister-type nebula like the Orion Nebula, we can actually map the object in three dimensions.

This mapping process begins with something called Ferland's mecha-

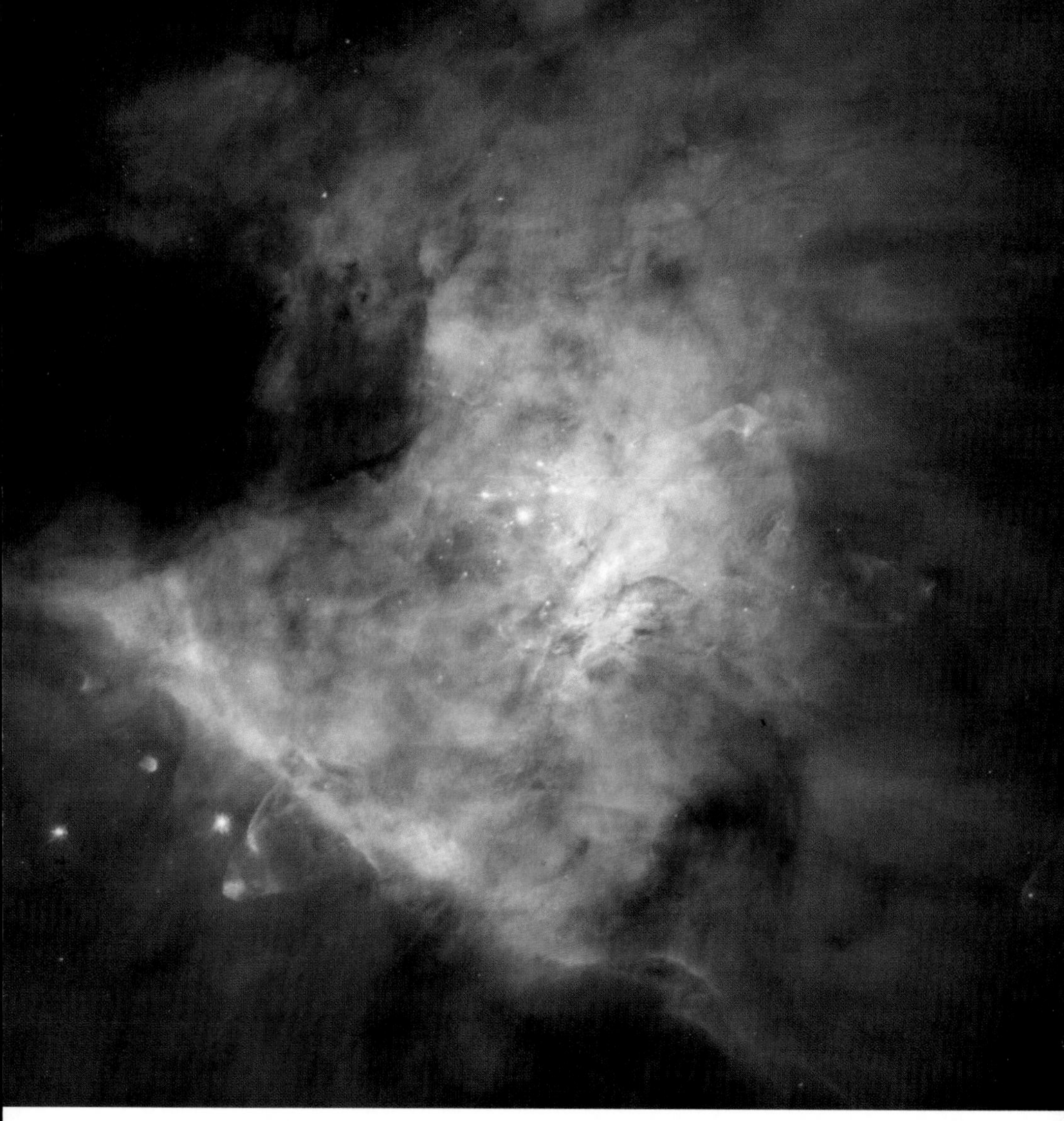

Figure 8.2a This and the accompanying image show the center of the Orion Nebula at an angular resolution comparable to the best ground-based observatories. This figure shows the nebula as it appears to an observer located in our solar system.

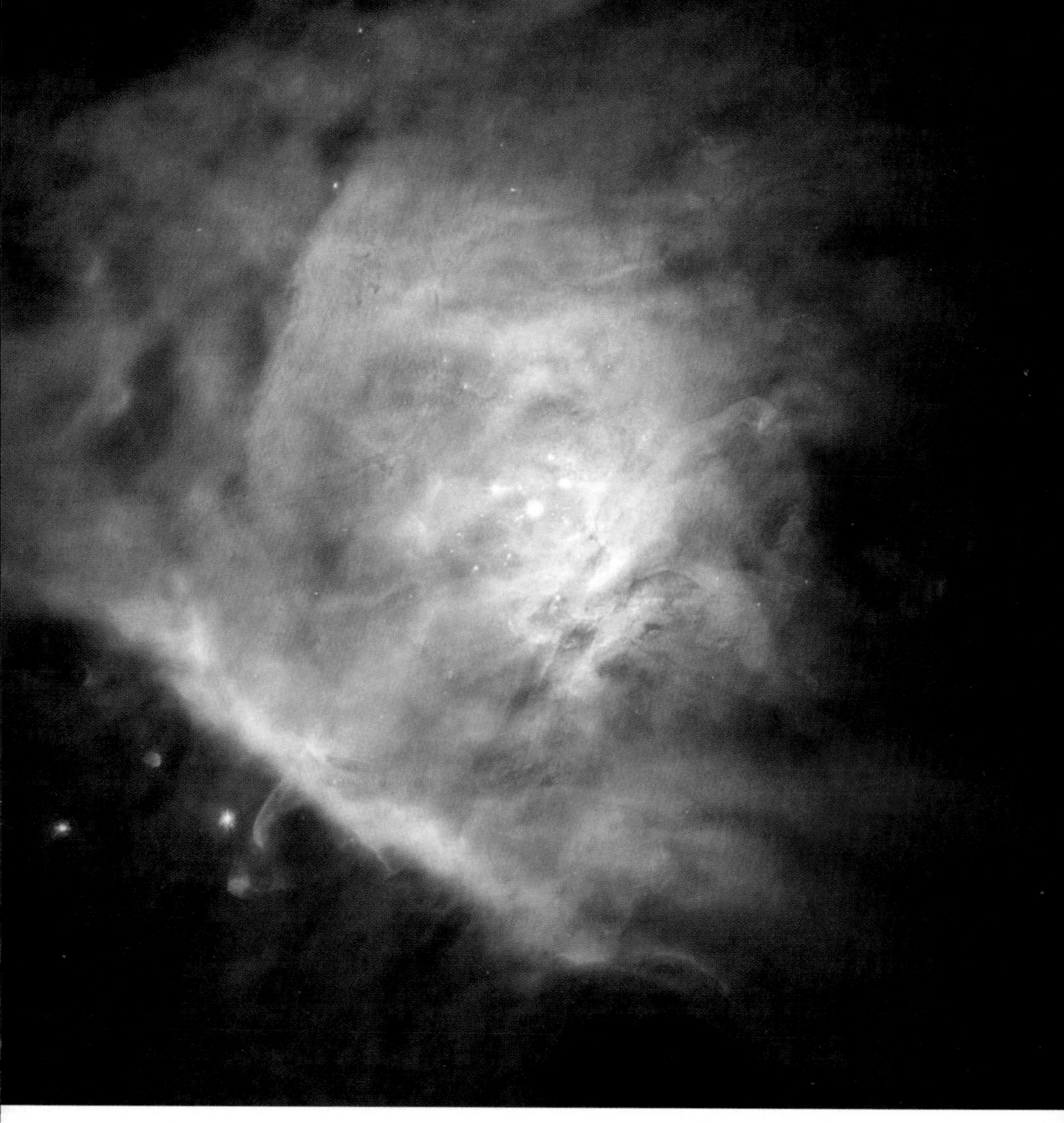

Figure 8.2b This figure shows how the nebula would appear if the extinction caused by the foreground veil of material did not occur. Extinction was removed by comparing Hubble and VLA radio images.

nism, named after Gary J. Ferland of the University of Kentucky. Ferland recognized in 1991 that the brightness of the surface of a blister-type nebula would be exactly proportional to the amount of ionizing ultraviolet radiation that it received. If there is a dominant source of this ultraviolet radiation, then the brightness of the nebula would be a direct measure of the distance between that point on the blister and the dominant star. Ferland's mechanism is a good first approximation. It explains why Orion becomes fainter farther from the hottest and most luminous star, Θ^1 Orionis-C (henceforth Theta1C). However, we can do even better by calculating the subtle differences caused by the light striking the surface from one direction and being observed from another, a task taken on by Zheng Wen of Rice University as part of his doctoral dissertation. The end-product of Zheng Wen's and my application of Ferland's mechanism is a three-dimensional map of the Orion Nebula. In this map the nebula lies at a distance of one light year beyond Theta1C. From that point it assumes an irregular concave shape, with the outer parts being closer to us than Theta1C in some directions, while to the east the surface actually dips away from us. The ionized blister (which is the visible nebula) is about 0.08 ly thick. The irregular shape that we see must be a result of the ionization front progressing faster or slower into the Orion Molecular Cloud depending on whether the local material was of lower or higher density. The Orion Nebula Cluster is centered on the Trapezium stars, being concentrated so much that in the center the density is about 20,000 times that of stars in the vicinity of the Sun.

In the last decade we have come to understand that we also see a foreground veil of material left over from the champagne phase. The gas in the veil forms absorption lines in the radio emission from the nebula and in the optical spectra of the cluster stars. The dust in this irregularly dense veil causes extinction of the light coming from the nebula. By comparing radio and optical observations of the nebula we can generate an extinction-corrected image of the nebula, i.e., derive a picture of what it would look like in the absence of this veil (see Figure 8.2b). Since the ultraviolet photons from Theta1C go out in all directions, the inside of this veil should be glowing, although much fainter than the nebula because it is much farther away from Theta1C.

Given a map of the three-dimensional structure of the nebula and

Figure 8.3 This is how the Orion Nebula and its associated cluster of stars would appear to an observer situated about one light year due east of the brightest star in the Trapezium. The main body of the nebula is the bright area in the lower middle of the image. The Bright Bar feature, which runs at a diagonal along the lower left of modern images of the nebula, is on the far left of this depiction. The overhead glow comes from the inside of the veil of material that lies in front of the nebula. This image is the result of combining Hubble images with our knowledge of the three-dimensional structure of the nebula and the star cluster (American Museum of Natural History and the San Diego Supercomputer Center).

---

armed with extinction-corrected images of the nebula, it was natural to combine the two. The combined data became part of a three-dimensional model of our galaxy. The Galaxy model was used in the opening planetarium show in 2000 at the Rose Science Center of the American Museum of Natural History in New York City. The three-dimensional model of the nebula and its surrounding region was put into a computer database in a fashion that allowed us to calculate its appearance from any position, including the inside (see Figure 8.3). As a sequence, this was used to give the impression of traveling through the nebula itself. Of course the producers of the animation had to make that trip occur quickly, which meant we had to ignore the relativity effects that would apply to an observer traveling a distance of about six light years in two minutes time.

The result of our collected data was a trip through the Orion Nebula as it is now (or at least the way it was 1,500 years ago when the light we now see was emitted). The nebula must have been quite different in the distant past when it first formed within the confines of the outer part of its molecular cloud. It will certainly look different in the future, when the ionization front has moved farther into the Orion Molecular Cloud. Of course the biggest change will occur when Theta1C begins to cool after using up its hydrogen fuel and the nebula becomes fainter and more compact. In a few tens of millions of years the nebula will no longer be visible and the stars in the cluster will begin to dissipate and spread out into the general field population of our Milky Way Galaxy.

CHAPTER NINE

Where Did All These Stars Come From?

WHEN we were kids the evasive answer to the question, "Where do babies come from?" was that the storks brought them, a legacy that is carried on in birth announcements but apparently nowhere else. For a long time our knowledge of how stars formed was little better than this. Once it was established in the 1930s that stars derive their energy from the burning of nuclear fuel, it was obvious that the hot massive stars were much younger than the age of the oldest stars in our galaxy, therefore stars must still be forming. A natural association with interstellar clouds was made because one could see that the young hot stars were concentrated along the spiral arms, just like interstellar gas and dust. Our knowledge remained at that level until radio astronomy came into its own after World War II. Interstellar neutral hydrogen had been detected by 1951 and in the 1960s radio astronomers began the detection of a wide variety of molecules in the interstellar medium that now numbers more than 100. It was quickly recognized that the molecules were largely concentrated in objects we now call molecular clouds. Once these were recognized, it became clear that they were probably the sites for forming new stars because they had masses of a hundred to a hundred thousand times the mass of our Sun. The great masses meant that they had plenty of material for forming even the most massive stars.

Conditions Inside a Cow

There are two primary competing forces in a molecular cloud, the outward force due to pressure and the inward force of gravitational attraction. For gravitational attraction to prevail, the force due to pressure must be overcome. Recall that pressure is produced by a combination of density and temperature. Clearly, if the pressure is too high, then a cloud cannot form stars. Fortunately, the temperatures inside a molecular cloud are incredibly cold, about 10 K. Such low temperatures have been produced on Earth in the physics laboratory, but only in very small volumes. In the case of a molecular cloud the sizes are characteristically light years.

The reason it is so cold in a molecular cloud is that it is incredibly dark, like the inside of a cow, to use another saying of my youth. In the absence of a local source of power, material is heated by the radiation it receives and absorbs. Recall that fine interstellar dust particles are inextricably mixed in with interstellar gas. This means that when a molecular cloud is formed, it will be a mixture of both. E. E. Barnard's photographs of the Milky Way showed that our galaxy contained numerous dark patches, almost devoid of stars. Detailed follow-up studies showed that these were not regions without stars in that direction; rather, that there was an obscuring cloud of material. The molecular clouds are simply an extension of these types of objects. The same characteristic that causes such a cloud to block out the light from the background stars will mean that starlight cannot get into the molecular cloud itself. A form of hydrostatic equilibrium (like Earth's atmosphere) exists, so that the giant molecular cloud gets denser and denser as one approaches its center. Increasing density implies more dust to block out the radiation, so the object quickly reaches its minimum temperature. The central densities can be very high, reaching many millions of molecules per cubic centimeter (recall that the average interstellar medium has about one atom per cubic centimeter).

The dust in a molecular cloud is a bit cooler than the gas, so as the atoms randomly strike the cool dust particles they stay there, much like dew forming on grass at night. The atoms temporarily trapped on the surface of a dust grain are concentrated much more than when they were freely moving components of the gas. This concentration makes it possible for the atoms to begin combining and forming molecules. Some of the molecules stick to the grains in a form of frost, while most of the others are knocked off when another atom arrives. Slowly the core of the molecular cloud builds up into a witch's brew of gaseous water, alcohol, formaldehyde, and even more complex molecules. However, most of the hydrogen will be tied up in a double molecule (H_2), and the next most common molecule (CO) will be formed from two of the most common heavy elements. It is easy for radio telescopes to detect the emission of CO and it is commonly used for observationally probing the inner structure of a molecular cloud. These clouds are highly turbulent and contain strong magnetic fields, both of which play a role in determining if a cloud can fragment and collapse. Molecular clouds also typically have multiple cores—tens to thousands of solar masses. These are the building blocks of stars and star clusters.

Turning Cold Clumps into Hot Stars

We really don't understand what triggers the formation of an individual star; however, a key necessary ingredient is that the region be of high gas density. We know that the molecular clouds are highly turbulent, which means that there will be a hierarchy of clumps and those clumps are the candidate prestellar clouds. A trigger is needed to get things started. This could be a collision with another cloud or the passage of a shock wave. The gravitational inward force will then prevail over the outward gas and turbulent pressure that had been sustaining the object against collapse, and the prestellar cloud will begin an inexorable collapse.

In the earliest phase of formation, gravitational collapse occurs almost unconstrained. Recall that the gravitational force between two bodies depends on the product of their masses and the inverse square of their separation. For a collapsing cloud the force will be greater the more massive the prestellar cloud. This means that more massive clouds will collapse much more rapidly than low-mass prestellar clouds. As the cloud approaches stellar dimensions, which can be over a hundred times the size of our Sun, then the gravitational collapse begins to compete with outward pressure from the hot gas of what must now be called a nascent star. The gas particles in the cloud gain energy as they fall toward their mutual center. According to Planck's law of radiation, they become self-luminous. The energy they emit is derived from their collapse, rather than nuclear fuel–burning. They certainly are stars already, but are given the name pre–Main Sequence stars to indicate that they have not reached the state in which the energy release from nuclear fuel–burning produces hydrostatic equilibrium and stops contraction.

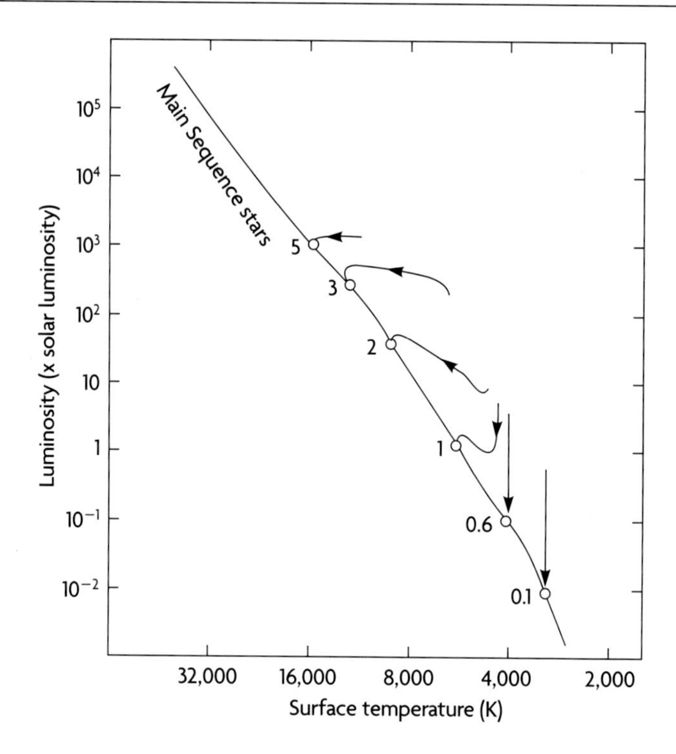

Figure 9.1 As protostars collapse under the effects of their own gravity their material is heated from this compression, and they become luminous. Where they will stop when they begin to burn hydrogen atoms into helium is determined by their mass. Numbers on the Main Sequence indicate the mass of stars as compared with our Sun. The pre-Main Sequence tracks of protostars of different masses are quite different.

The track of these pre–Main Sequence stars can be theoretically calculated. Sample results are shown in Figure 9.1, where the positions traced by pre–Main Sequence stars of different masses are depicted. Since the more massive pre–Main Sequence stars will collapse to the Main Sequence in shorter times than a low-mass star and we know the tracks that individual pre–Main Sequence stars take, we can do something like (but the opposite of) examining the H-R diagram for a young cluster. A cluster of stars is composed of stars of a variety of masses. If the cluster is rich enough to have any rare massive stars, they will have collapsed rapidly to the Main Sequence in a period of less than 100,000 years. We'll find them still on the Main Sequence unless the cluster is just old enough that they've begun to evolve through depletion of their hydrogen fuel. Most of the stars will still be raining down on the Main Sequence, with the higher-mass pre–Main Sequence stars closer to their final positions than their lower-mass brethren of the same age. This means that a cluster of pre–Main Sequence stars will be spread out differently in the H-R diagram in an age-dependent way, as shown in Figure 9.2. The situation is similar to the determination of the age of an evolved cluster, in that one must look at both the point where one finds stars on the Main Sequence and the distribution of stars that are above that line. If we observe the distribution on an H-R diagram of stars in a young cluster, we will find some on the Main Sequence and the cooler stars distributed above the Main Sequence. The most massive stars on the Main Sequence won't tell you anything about the age except that the cluster is older than about 100,000 years but younger than the time it takes the Main Sequence stars to burn up the supply of hydrogen in their cores. However, comparison of the location of the cool stars with the timelines shown in Figure 9.2 will tell us the age of this group of stars.

Rotation Makes Things More Complex

The formation of an individual star is more complex than the simple collapse of a small gas cloud because of the property called angular momentum. Linear momentum is the property that causes a moving object to want to keep going in a straight line (the origin of centrifugal force, dis-

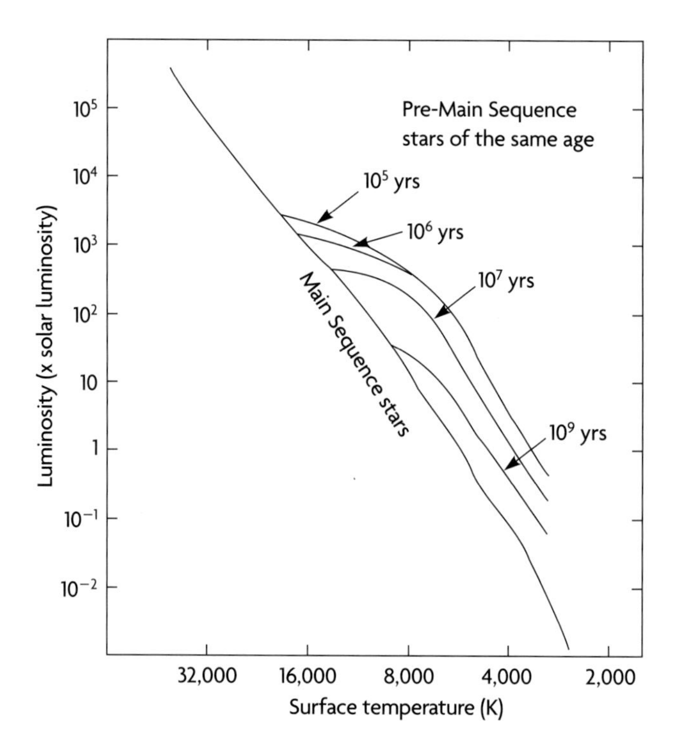

Figure 9.2 Since the massive stars collapse to the Main Sequence faster, a group of protostars that all were formed at the same time will occupy a curved line in the H-R diagram. In each case the most massive stars have already reached the Main Sequence. As the cluster of stars becomes older, a larger fraction of the stars will have reached the Main Sequence. One can determine the age of a cluster of stars and protostars by comparing their H-R diagram with these theoretical predictions.

cussed in Chapter 6). Angular momentum is the analogous property of a spinning object wanting to continue rotating. The amount of the angular momentum is determined by the mass of the object, the rate of rotation, and the distribution of the mass. In general all of the small clouds that begin to collapse toward becoming a pre–Main Sequence star will spin to some degree.

As a pre–Main Sequence star collapses, the material is compressed. In the absence of some force that acts as a brake, the amount of angular momentum will remain constant. Since the material is now more compact, this conservation means that the object will spin faster as it gets smaller. You can see this conservation of angular momentum preserved when you watch a spinning figure skater speed up simply by pulling in their arms. Eventually a point will be reached where the conditions mimic those shown in Figure 6.1: the outward centrifugal force from the spinning will be equal to the gravitational inward force, so the material will be left in a condition of quasi-equilibrium—it can collapse no farther. This won't apply to the entire pre–Main Sequence star, rather, only to the portions near the equator. The end-product is that material is left behind in a disk of gas and dust while the remainder becomes a nearly spherical star. Our solar system is an example of this mechanism. The planets (which formed from the material left behind) are in a thin plane called the ecliptic, and 98 percent of the angular momentum of the solar system is found in the planets, even though they contain but 0.13 percent of the mass.

Once the disk is formed new processes begin to occur, because now the density is so high that things like viscous drag of the material become important. Once viscosity takes a role, some of the material loses its angular momentum and drops out of orbit, falling into the star. In fact, material seems to continue to gravitationally fall down onto the inner disk, from which it is then fed onto the nascent star after having lost its angular momentum. Stars can even double their original masses through this process.

Stars, Brown Dwarfs, and Planets

Hydrogen is the most abundant element in the universe and the most abundant source of fuel for nuclear fuel–burning. Recall the discussion in Chapter 6 and the fact that the central temperature and density of a star depend critically upon the amount of mass of the star. It should be no surprise that stars of less than 8 percent of the solar mass are never hot and dense enough to burn hydrogen into helium. This means that the true Main Sequence of stars on the H-R diagram begins with objects of 8 percent the mass of the Sun.

We also need to consider the burning of deuterium into helium. Deuterium is a variation of the hydrogen atom. Its nucleus has an extra particle, a neutron. This is a rare, so-called isotopic form and is generally only about 0.0015 percent the abundance of hydrogen. However, deuterium is much easier to convert to helium through nuclear fuel–burning. In fact, this is the form of hydrogen that is used in attempts here on Earth to harness nuclear fuel–burning as a source of low-cost energy. Since there is a lot of seawater, the idea is to separate out this rare but desirable atom, which can be done cheaply. Since deuterium is easier to convert, conditions for burning it are found in objects down to 1.3 percent of the solar mass. Objects in the mass range of 1.3–8 percent of the solar mass are called brown dwarfs. Their surfaces are much colder than ordinary stars, with temperatures down to 1,400 K.

Objects formed with less than 1.3 percent of the solar mass (13 Jupiter masses) will never burn nuclear fuel except for the very rare and reactive element lithium. They can be defined as planets because any luminosity is simply due to their heating through gravitational contraction. The exact name that is applied really depends upon how they are formed. If formed in a circumstellar disk, they are called planets. An object formed through processes similar to those of star formation can be quite different from the same mass object formed in a disk around a pre–Main Sequence star. The former have been called rogue planets, free-floating planets, and even planetars, something between a planet and a star.

How Many Small Stars? How Many Large?

We actually understand the last phase of a star's formation rather well, because it has become self-luminous and observable with high-resolution cameras and spectrographs in optical and infrared radiation. Likewise, we know a lot about what is happening in the molecular cloud. What we don't understand very well is just how the onset of contraction is triggered. Before it happens we only see the material in the cloud, and once the contraction starts the star very quickly falls to the point at which the contraction lines begin in Figure 9.2. Our level of knowledge of these earliest phases of contraction is at the level of "the stork brings them."

Whatever is happening, it seems to be a process that takes a lot of ma-

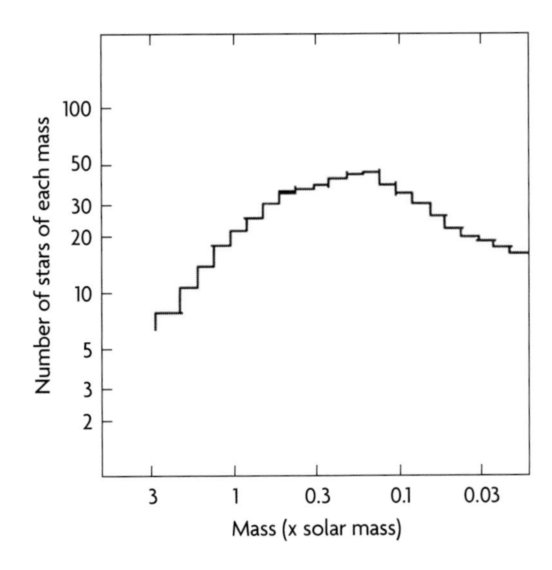

Figure 9.3 How many stars are formed with particular masses reflects the still-mysterious details of the fragmentation of molecular clouds into protostars. The most massive stars are quite rare, whereas the most common star has about 25 percent the mass of our Sun. The stars with masses less than about 8 percent of the Sun's mass are brown dwarfs, stars that will never be hot enough to burn ordinary hydrogen into helium.

terial. Most stars form in clusters from giant molecular clouds. Stars aren't born individually, but in litters. In the case of the Orion Nebula Cluster, there are about 3,500 stars, containing a total mass of about 1,500 solar masses. The fragmentation process isn't instantaneous, but is concentrated in a brief period of less than a million years. More massive stars are formed near the middle of the molecular cloud, since molecular clouds are most dense there.

Although we don't understand the process of this early fragmentation now, there is hope that we will. This is because there seems to be a very similar distribution of stars of different mass regardless of the exact region of origin. This distribution is known as the Initial Mass Function. In Figure 9.3 we see the Initial Mass Function for the Orion Nebula Cluster. At the highest (hottest) end there are only a few stars in the sample, whereas on the lowest (coolest) end the limit is observational; the stars are too faint to be seen against the glow of the nebula. The middle range of masses is well defined and we see the most common stars are about 0.2 times the Sun's mass. However, in terms of brightness, the most massive (hottest) stars are the most important. Although there are only one-tenth as many stars of 3 solar mass as of 0.3 solar mass, the former are a thousand times more luminous and collectively contribute a hundred times the amount of radiation as the latter. Since the structure of the ionized nebula is determined by the highest-energy ultraviolet radiation, only a few stars are important. In fact, Theta1C contributes twenty times as much ionizing radiation as the next brightest star in the vicinity. At the other end (the lowest-mass stars), stars below 8 percent solar mass are brown dwarf stars and will never be hot enough to burn their hydrogen fuel. Figure 9.3 stops just short of including any planetars. Very recent work indicates that there are such objects in Orion, so I will discuss them in Chapter 11. Since the Initial Mass Function is very similar for all well-studied star clusters, it is probably true that what we see happening in Orion is characteristic of the rest of our galaxy.

CHAPTER TEN

The Hubble Space Telescope

On April 24, 1990, the Hubble Space Telescope was launched into Earth orbit. It is the product of twenty years of intensive and expensive planning and engineering. It has produced unique science from the beginning, survived an early period of crisis, and become the best-known astronomical observatory in history. In this chapter I'll explain why this observatory was built and give a brief insider's history of its development.

Even a Clear Night Is Not Enough

In Chapter 5 I discussed why our atmosphere limits what the earth-bound astronomer can view. We can observe radiation that comes through the optical and radio windows. Radio astronomers are fortunate in that they can observe even through cloudy conditions, although they usually record an image that is limited in quality by the size of the antenna. The optical window requires a clear sky and usually produces an image of better quality than the radio window.

The quality of an image is indicated by something called the angular resolution. This indicates how close two sharp sources can be brought toward each other and still appear as two sources. Another related measure of the quality of an image is the width of a source that is extremely small. Images formed by telescopes are brightest in the middle, then fade quickly as you look away from this center. It is as if you were looking down on a conical mountain peak. The width of an image is usually measured by something called the full width at half maximum, which is the diameter of the image where it is one-half of the maximum brightness. For the usual case this width is very close to the angular resolution. In Chapter 4 we saw that for radio telescopes the angular resolution depended on the wavelength divided by the size of the telescope, the so-called diffraction limit. The same will be true for accurately constructed optical telescopes. This means that the 200-inch Palomar giant telescope should be capable of forming an image only 0.03 arcsec in size, if it were perfect. The image it forms is not perfect, however, because the atmosphere limits the image quality to much worse than 0.03 arcsec.

What's an arcsec? It is a unit of measure for very small angles. There are 3,600 arcsec in one degree. This means that the Sun and Moon each subtend an angle of about 1,800 arcsec. The human eye can resolve an angle of about 50–100 arcsec.

The visual size of a star image for a ground-based telescope is about 1 arcsec. This is about the quality of the image formed by a 4-inch telescope. For telescopes much larger than this, the quality of the image is constant at about 1 arcsec and is determined by Earth's atmosphere, rather than the telescope. This is a very serious limitation indeed!

The culprit is the fine-scale fluctuations in the temperature of Earth's atmosphere. Each packet of turbulent air of a different temperature bends the incoming starlight very slightly. When the telescope is so large that you are looking out through multiple packets of air, then the image formed by a large telescope is a combination of light bent in many directions. The image is stationary, but fuzzy.

Early seventeenth-century telescopes were small and of poor optical quality. Their angular resolution was determined by their size and quality, not the atmosphere. However, it has been feasible to build larger and higher quality telescopes since the early eighteenth century. This has meant that for three centuries it has been Earth's atmosphere rather than the telescope size that has determined how sharp an image of a celestial object we can study. An observer with a 200-inch telescope is capable of viewing no sharper an image than a well-equipped amateur observer with a 10-inch instrument; however, a bigger telescope gathers more light and forms a brighter image.

The quality of an image is also dependent on time, site, and wavelength. Even at a single location the image quality can vary during the night, depending on things like cooling after daytime heating and wind conditions. Sites vary enormously in their average image. Palomar Observatory has a long-term average image of almost 2 arcsec and Mauna Kea in Hawaii averages slightly better than 1 arcsec. In general the image improves (becomes smaller) with increasing wavelength, because that type of light is bent by a smaller amount at a fixed temperature difference. This improvement continues until the ratio of wavelength to telescope size becomes large. Then it becomes limited by diffraction rather than by atmosphere. One can finesse the observing process by picking the optimal site, observing at the optimal wavelength, and choosing the optimal time. Even though there are promising methods approaching maturity that circumvent the limitations of atmospheric blurring, the fact remains that Earth's atmosphere is a fundamental constraint on how sharp an image one can form and therefore how well we can observe the universe.

Circumventing Atmospheric Turbulence

The basic limitation of atmospheric turbulence has long been recognized. This is part of the reason George Ellery Hale of Yerkes Observatory built the 60-inch reflector on Mt. Wilson in southern California rather than in Wisconsin, where the Yerkes 40-inch was located. Since then astronomers have sought out the sites with the clearest skies and sharpest images. The culmination of this process has been the crowding of thirteen telescopes onto the summit of Mauna Kea in Hawaii. Even under Mauna Kea's superb conditions, atmospheric turbulence has been a serious problem. The cleanest long-term solution, recognized for over half a century, is to put a significantly large telescope outside of Earth's atmosphere.

This solution was slow to be recognized. As a post-doctoral fellow at the Mt. Wilson and Palomar Observatories I was able to observe every month at one place or the other. Not all nights were clear, of course, and this gave me the chance to read through the file of crank letters. The crank file is a common institution at many observatories, typically filled with unsolicited mail from writers with ideas they want to share with professional scientists. This mail is often written by people without a science background. A common theme is a "theory of everything," wherein the writers claim to be able to explain everything by a few new laws and ways of looking at things. The urge to explain the universe as simply and reductively as possible seems to be part of the human condition. Not all letters are quite this implausible. One of the letters in the Mt. Wilson file was sent during the long construction of the 200-inch telescope. In it the writer argued that locating the instrument on top of Palomar Mountain was the wrong thing to do. Rather, the telescope should be flown high in Earth's atmosphere suspended by a gigantic balloon, thus being freed of the blurring effects of turbulence. Although the plan was then impractical and the letter ended up in the crank file, it was ahead of its time in an important way.

One of the most dynamic astronomy groups in the United States is located at Princeton University. Its modern form began in 1947 with the appointment of Lyman Spitzer, then thirty-three years old, as professor and head of the department (Figure 10.1). At the same time Spitzer convinced Martin Schwarzschild to leave Columbia University and join him. Spitzer had long been convinced of the multiple opportunities offered by observing from above the atmosphere and the Princeton astronomers and engineers undertook a series of projects that were directly related to building the Hubble. The first of these involved flying high-quality telescopes from balloons above most of Earth's atmosphere, as advocated by that letter in the Mt. Wilson crank file! By the summer of 1959 the first of these instruments was flying. Dubbed Stratoscope I, it flew a 12-inch telescope and obtained images of the Sun at an unprecedented resolution. This project was succeeded by the much more ambitious Stratoscope II project, which flew a 36-inch telescope and was intended to take long-exposure images of faint planetary and stellar objects. This program was plagued by many difficulties and only a few flights could be

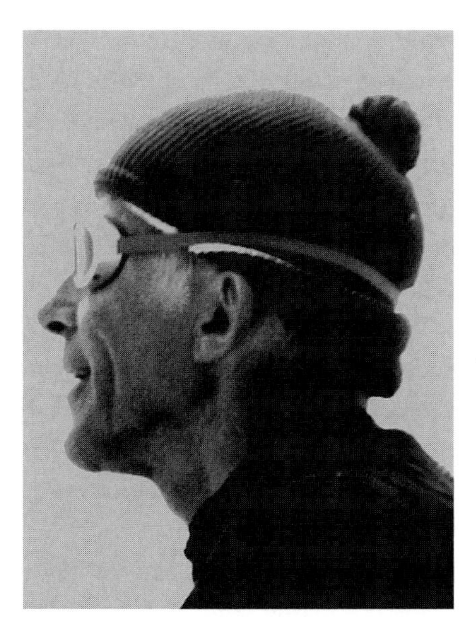

Figure 10.1 Lyman Spitzer is considered by many to be the father of the Hubble Space Telescope. He was a gifted astrophysicist and leader in the development of rocket and satellite instruments. He was also an avid rock climber and mountaineer. This photograph, by the author, was taken just below the summit face of Mount Robert M. Petrie during the first ascent in July 1967.

considered successful from a scientific point of view, but the experience gained was invaluable. Some of these same Princeton scientists and engineers pursued technology projects that could be used in an image-producing orbital telescope, while others became involved with the highly successful third *OAO* spacecraft, named *Copernicus,* which obtained high-resolution spectra in the ultraviolet in the early 1970s. The Princeton group was not aware that at the same time very similar imaging technology was being pursued for very different purposes.

National security needs demand as much information as possible about the nation's perceived enemies. One method of spying is to get pictures of what is happening. Even in the Civil War, aerial surveillance was an important source of information. Those balloon-borne Union observers were mighty vulnerable, as were the World War I aircraft, but they were the best means of surveillance available. By World War II stripped-down, unarmed versions of military aircraft could fly higher than their pursuers and operated with relative impunity. This evolution of technology continued with development of the U-2 single-engine reconnaissance aircraft, which secretly roamed above the Soviet defense systems during the Cold War until one of them was brought down in 1960 and its pilot captured, much to the discomfiture of the United States. Clearly, space satellites offered a safe alternative for pursuing these same goals. Telescopes of increasing aperture were built and operated on these satellites, always pointed down. Many were in low Earth orbit, in order to be closer to the objects being imaged, and paid the price of having short lifetimes as a result of drag from the upper reaches of the atmosphere. Early satellites packaged the undeveloped film in reentry pods that were then snatched in midair by their parachute shrouds. Later ones used electronic detectors that could send the signal with the image information down to ground stations. Astronomers wanted this very technology for looking up into space, whereas the reconnaissance satellites were intended strictly to look down at Earth's surface.

Different needs produced different telescopes. Astronomical telescopes make long exposures of objects fixed on one point in the sky, but reconnaissance telescopes make short exposures of quickly moving objects. Although the needs of the two parties were quite different, there were many elements in common, for example, the ability to generate a highly precise lightweight mirror. U.S. national defense technology was being developed in secret, but in parallel with the Princeton activity and provided the experience base that was used for the Hubble.

This Idea Has Been Around a Long Time

The idea of something like the Hubble has been around about as long as realistic visions of space flight. In his book *By Rocket into Planetary Space* Hermann Oberth, a German rocket pioneer, wrote about the advantages of observing with a large telescope in Earth orbit. This vision was well ahead of its day since in 1923 even modern liquid-fueled rockets were still a dream. As described in Chapter 5, that technology was developed in the 1930s and the early 1940s by Werner von Braun and his German team, with the remaining V-2 rockets eventually finding a home in the United States. Although the American Robert Goddard had pioneered liquid-fuel rockets, his work was not widely known or appreciated. Therefore, it was not until the United States possessed the remaining V-2s that Americans began to see the potential of space flight and the government funded a project to articulate the possible uses of spacecraft. Lyman Spitzer, then at Yale University, participated in this project. He clearly posited the advantage of an orbiting observatory. The resulting report was a classified document, so it was not widely known among astronomers. However, when NASA was created in 1958 it began forming long-term goals. In 1962 the Space Science Board of the National Academy of Sciences recommended a Hubble-like observatory as a natural long-term goal for NASA. This was followed (1966) by a committee study chaired by Spitzer, whose report ("Scientific Uses of the Large Space Telescope," 1969) articulated the benefits of a Hubble-class space observatory. "Large Space Telescope," the name used in the early years, was revised to "Space Telescope." The final name "Hubble Space Telescope" was decided upon late in the construction phase. A continuous stream of technology projects and the OAOs, last launched in 1972, gave credence to the project (as did the reconnaissance satellite work, then still unknown to almost all astronomers and most within NASA). In 1971 Nancy Roman, who headed NASA's astronomy program for several decades, convened a group of scientists (I was one of them) to advise NASA regarding the feasibility of the Hubble supplementing a more formal technical study. The aperture considered was the round number of three meters that had been proposed by the Spitzer committee. In 1972 the project moved into preliminary design and I relocated from Yerkes Observatory to Huntsville, Alabama to serve as project scientist. Nancy Roman was the program scientist at NASA headquarters, responsible for representing the Hubble within the upper echelon of NASA management.

An Insider's History of Building the Hubble

It is significant that NASA picked a project scientist who was not already an employee, and that I would be working at the Marshall Space Flight Center, rather than the Goddard Space Flight Center, which had been the center of most of NASA's astronomy activity. It was a wise political move because the Hubble was not then widely supported among American astronomers. Hiring the head of a highly ranked astronomy program would lend credibility to the project and make it easier to attract other major figures. The common attitude among ground-based astronomers was that the Hubble was a fine goal, but that a host of large ground telescopes built with half the money would yield more scientific activity. What these astronomers didn't take into account was that in the real world things don't work that way. Failing to build the Hubble (within the NASA budget) would not produce money in the NSF budget for enhancing ground-based astronomy. By 1973 we had assembled a senior group of scientists (the Space Telescope Science Working Group) from major U.S. and European institutions. The collective knowledge of the members and the prestige of their backgrounds produced community-wide support for the project and opened doors in congressional offices when it became necessary to lobby to keep the Hubble project alive.

NASA field centers answer to NASA headquarters, but each has its own history and specialty. At some level they compete with one another. Marshall had a strong engineering reputation from the development of the Saturn rockets. Werner von Braun and his team, settled in Huntsville at the end of the American V-2 program, worked within the Redstone Arsenal organization there to develop new rockets, quickly launching the first U.S. satellite soon after the launch of *Sputnik.* When NASA was formed von Braun's team became the nucleus of Marshall, the co-located NASA facility. By the late 1970s it was clear that the end of the *Apollo* program was near and that the need for Marshall's primary product (rockets) was going to diminish. Von Braun began looking for new projects, including development of large scientific payloads. With the help of his chief scientist, Ernst Stuhlinger, he convened a group of three (Herbert Friedman, Peter Meyer, and me) to advise him about which projects to pursue. I had not been deeply involved with space projects since I was a graduate student, but was a member of NASA's Astronomy Missions Board. Herbert Friedman had been closely involved with space projects for years, beginning with the American V-2 program. Peter Meyer had also been involved with space projects in pursuit of his specialty of cosmic ray physics. Meyer had been a graduate student of Stuhlinger in Germany, immediately before Stuhlinger was drafted into the German army. Until the end of World War II, Meyer remained in the protection of Christian families because of his Jewish heritage. Friedman, Meyer, and I recommended that Marshall pursue several programs, one of which was the Hubble.

The lack of a large established body of astronomers at Marshall made it easy to draw in appropriate outsiders and make the Hubble program responsive to their needs. During the next decade we worked closely with both Marshall and contractor engineers to identify scientific requirements and turn them into engineering requirements. The Hubble was given the go-ahead for construction in 1977. We selected a group of scientists to develop individual scientific instruments and work on the telescope. It is poignant that none of the Princeton astronomers, whose work led indirectly to the Hubble project, were chosen in that competition. Their intensive involvement in earlier projects proved disadvantageous; they had tied themselves to certain technological approaches that quickly became outdated.

When it was time to address the question of how the Hubble would be operated, things became difficult. Goddard believed that they should continue the role they had fulfilled in the *OAO* program and were planning for the *IUE* (which later proved highly successful). The outside scientists wanted more control and looked to the highly successful models of the national optical and radio observatories, run by autonomous scientific organizations but with NSF funding. NASA, however, was accustomed to controlling the utilization of its scientific satellites and there was great resistance to the idea of an independent center. In the end, the same organization that runs the national optical observatories was selected to establish the Space Telescope Science Institute near Johns Hopkins University. Scientific users of the Hubble work with this institution, whereas Goddard is the lead center for the actual operation of the spacecraft and development of each of the essential servicing missions, which occur at about three-year intervals.

At about the same time, NASA entered into an agreement with the ESA that called for the Europeans to provide the solar arrays that would power the spacecraft and one of the initial science instruments. They also were to provide part of the staff for the Space Telescope Science Institute.

Their financial share of the burden, which continues at the writing of this book, is 15 percent of the total. ESA member-nation scientists receive a comparable proportion of the observing time.

I stepped down as project scientist in 1983, after eleven years. I had relocated to Rice University the year before, when we thought the Hubble would be launched shortly and it wasn't necessary to have a resident full-time project scientist. I had to make a decision: should I return to being a scholar and research scientist or finish my career as a NASA manager? Even though I thoroughly enjoyed working for NASA and had taken on the additional role of associate director for science at Marshall, management is not why I became a scientist. In 1983 it was obvious that there would be a significant launch delay because there had been engineering and financial problems. A new project scientist was named and I stayed on as a member of the Science Working Group, becoming head of the team responsible for issues involving the observatory as a whole. This put me in a position to return to practicing science and to prepare to become an effective user of the Hubble, always my goal. However, there was an obstacle to being one of the initial users. The initial users were persons competitively selected in 1977. As a member of the structure that had managed that selection, I was excluded. I had resigned from a tenured professorship at the University of Chicago, had worked eleven years on the project, but still had no guaranteed use of the Hubble telescope, whereas many of my colleagues did.

Then occurred the greatest surprise of my life. At a reception during one of the Science Working Group meetings held at Goddard I was presented with an astronomical photograph and signed plaque. The plaque indicated that all of the other seventy scientists selected to use the Hubble had surrendered a fraction of their time, so that I would be given an equal share of observing time. This was a boon in the most valuable of scientific currencies! The NASA managers in headquarters financially backed this independent action by the scientists. This was about 42 orbits (a little more than three days) of observing time. In consideration of the fact that Lyman Spitzer had been so deeply involved with creating the Hubble, but his camera proposal had not been selected as one of the flight instruments, I then divided my share with him, leaving him great

Figure 10.2 When we built the Hubble Space Telescope we considered it worth the investment of large amounts of money and decades of effort. No one is disappointed (Space Telescope Science Institute and NASA).

autonomy about how he actually used the time. Having guaranteed observing time was a great pleasure; being able to share that time with Lyman was a great satisfaction.

Missteps

The shakedown period immediately after launch of a spacecraft as complex as the Hubble is always demanding and marked by some problems, but nothing had prepared us for what happened. In spite of all the best efforts of the NASA and contractor personnel, the image formed by the telescope was not as we had expected, and the spacecraft was vibrating as it passed into and out of Earth's shadow. The former problem presented a fundamental constraint on what we could do with the Hubble. The latter problem was almost completely eliminated by changing some of the on-board computer programs. However, they both required major fixes during the first refurbishment mission.

The optical problem was quickly isolated to the primary mirror. It had been manufactured with great precision. This meant that the error lay with the test procedure used. The test procedure was designed to monitor the shape of the mirror as it was polished to the desired specifications. We knew that the mirror actually exceeded the minimum contract specifications in terms of deviations from the optimum surface as determined by the tests. The problem was that it was the wrong shape.

How could this happen? To understand this you must understand how the shape of a mirror is established. One type of mirror—a sphere—is very simple to test. In the nineteenth century Jean Bernard Foucault demonstrated a simple method of testing spherical mirrors. If you put an artificial star at the center of a spherical mirror, an image of that star would be formed adjacent to the light source. If you then take a sharp edge and occlude that image, the primary mirror would appear as a bright, flat disk. The task of the optician was then to polish away the apparently high and low spots until the mirror looked flat in this test. The same procedure is used today, except that one uses a more exact way of measuring the apparent flatness called interferometry (measuring the interference of light waves). Other shapes of mirrors will not look flat in the Foucault test. The optical design used for the Hubble telescope was one developed in the early twentieth century by George Willis Ritchey and Henri Chretien. The Ritchey-Chretien design called for a primary mirror of very complex shape, but one that was well specified.

The Hubble primary mirror was tested by use of a device that takes the light of an artificial star and modifies it so that when you look at the primary mirror, it appears flat when the primary mirror is exactly the right shape. The complex test device therefore reduced the optical tests for the Hubble's mirror to the equivalent of a simple test for a spherical mirror. Since we knew that the test results did show an apparently flat surface, the problem lay in the test device itself. This device had three lenses (two mirrors and one refractive lens), each of which had been manufactured with great precision, which narrowed the search down to whether or not these three lenses of the test device had been put into the correct relative positions. It was found that one of them was out of position by 1.3 mm, an amount a thousand times larger than the expected uncertainty.

The incorrect placement of the lens occurred through a bizarre error. An optical device was used to determine the location of an extremely precise spacing rod. In order to look at the very center of the polished end of the rod, a cap was fitted over the end. This cap had a small hole in its center and the measuring device was pointed at the rod end through this hole. The rest of the cap was coated with black paint, so that as the technician moved the optical measuring device around, looking for the hole, light would only be seen through the hole. However, the paint had chipped at the edge of the hole, so that the technician locked onto the strong signal coming from the cap, rather than the end of the rod. When he then put the lens into position, he found that he had to add shims to bring it into position. This peculiarity should have raised a red flag, but it was only casually documented, and not reported to the Hubble project managers at Marshall.

The misplacement of the refractive test lens meant that although the primary mirror looked perfect, it was actually the wrong shape. The deviation was small, only 0.0003 mm, but when you are trying to produce exquisitely small images, it is serious. After we determined the actual shape of the mirror, which turned out to be the same shape we inferred from in-orbit measurement of star images, the path to correcting the situation was clear. We needed to add an additional optical correction so that the cameras mounted at the focus of the Hubble could form sharp images. This was accomplished with a simple device installed during the first servicing mission of the Hubble in December 1993. One of the scientific instruments was removed from the spacecraft and replaced with the correcting device, which modified the light being fed to the cameras and spectrographs. The results were superb; the corrected images were all we had ever expected. The most frequently used camera, the Wide Field and Planetary Camera, had been slated for replacement with an improved version on the first servicing mission, and the relay optics within that camera were simply modified to accommodate the unexpected but now well-known shape of the primary mirror. No one would have wished for the problem, but in the end the Hubble is working just as we had hoped. Each scientific instrument since this refurbishment mission has its optical corrections applied internally.

The Hubble's vibration when entering and exiting Earth's shadow resulted from instability in the arms projecting from the Hubble that hold the solar arrays, which provide the electrical power to run the observatory. These arms undergo a big change of temperature as the Hubble goes into and out of shadow. The thermal design of these arms was inadequate, hence the arm adjusted its shape more than expected. Crucially, this adjustment did not occur smoothly, rather, it changed in a fashion called stick-and-slip. These little jumps would then shake the entire spacecraft by about 1 arcsec, which was intolerable since the expected image size was about 0.07 arcsec. A work-around solution was found by simply not making observations for the few minutes immediately after the transition to and from Earth's shadow and by changing the software within the pointing and control system. Redesigned solar arrays, installed during the first servicing mission, then solved the problem. Higher-power solar arrays replaced these during the fourth servicing mission in 2002.

Before the first servicing mission, images from Hubble had a core just as sharp as we ever expected, but surrounded by a halo of light. The Hubble had so many different programs it could address, it was not difficult to identify from among them a full slate of short-exposure observations that were not seriously affected by the peculiar nature of the image, although the exposure times necessary were longer. The Hubble produced a lot of unique and valuable science even before the first servicing mission.

It was interesting to track how the media covered the Hubble. Prior to launch there was a media circus of pleasurable anticipation that we participated in, often with no small amount of hyperbole in our enthusiasm. Things came crashing down with the public announcement of the optical problems and the Hubble became the whipping boy of critics of big projects and of NASA. As the scientific results filtered in, media stories often began, "The crippled Hubble has shown . . ." This then evolved to leads of "In spite of the flawed optics, . . ." Then there was a period when the flawed optics were obligatorily mentioned at the end of every article about the Hubble, until finally all reference to the problems was dropped. This evolution perhaps reflected media acceptance of the fact that the Hubble still produced good science in spite of a faulty mirror or vibrations.

How could the error in the test device for the mirror have slipped through undetected? Like all accidents, there were many factors. The critical issue was that there were no formal cross-checks. The Hubble program was seriously constrained by costs, so NASA management accepted the development plan of the optical contractors. Their plan assumed that things would go well as long as the individual steps were done right. But one of those steps, the erroneous alignment of the measuring device on the flaked portion of the cap over the end of the spacing rod, was compounded by the failure to recognize the significance of needing to add shims to the test lens. Later tests, which employed additional optical systems to look at the primary directly or through the test device did show that something was wrong with the mirror. These deviations were always dismissed as being due to the fact that those testing devices had not been manufactured to the same high specifications as the primary test device. The Hubble project was operating under enormous time and financial constraints, so the easy explanations seemed acceptable.

There is another dimension to how the problem slipped through. One large component of the non-NASA space program is the building of reconnaissance satellites that employ telescopes to image Earth. If the Hubble had been the first instrument to use significantly sized precise mirrors made and tested by these techniques, more attention would have been given to them. The manufacturer of the optics had built an extremely precise 1.5-m mirror as proof of their ability. They were so confident with the testing approach used that they did not review the anomalous need for an extra spacer.

The Hubble is not simply a reconnaissance telescope that looks up instead of down. The mirror technology may be the same, but other elements are different. In particular, reconnaissance satellites must have the ability to move quickly as Earth rushes past below. The Hubble has to remain pointed in space with very great precision for long periods of time, hence the pointing and control systems are very different. The heart of the Hubble's pointing system is a set of three instruments located on the edge of the field of view of the telescope. The Fine Guidance Sensors look at stars, track how they are starting to move, and provide a signal to the spacecraft that allows correction for this motion. A need unique to Hubble, the Fine Guidance System design was difficult. While the Hubble's primary mirror was being fabricated, we learned that the original Fine Guidance System design would not work well enough to meet our pointing requirements. Without good pointing we couldn't utilize those good images. Solving this problem became the top priority of the optical engineers and scientists involved. In the end the new Fine Guidance System design has worked very well, but in that period of crisis, attention had been diverted away from the few red flags that appeared within the parts of the Hubble development that were considered straightforward.

Another factor was the constraint on manpower put on the Hubble project by an agreement with the Department of Defense. In order to limit the number of people who would have access to necessary classified information and facilities, manpower levels were established that were much lower than normal. At all levels of management within NASA, engineers and quality-control staff were required to supervise and monitor much more technical work than normal, so that keeping a close eye on the manufacturer was not possible. Only one full-time on-site NASA quality-control person was assigned to provide oversight for the entire telescope portion of the project. His security clearances were provided so late in the program that he was unable to understand the capabilities and challenges in building a large, precise space telescope mirror.

Like birth pains, these difficulties have faded because of the enormous success of the Hubble. By this time most of the original scientific instruments have been replaced with higher-performance devices, just as we had planned when designing this first man-maintained orbiting observatory. A final servicing mission is planned and we have reason to hope that the Hubble will reach two decades of operation, even though the original design lifetime was fifteen years.

There are several books presenting the history of the Hubble project. Some are superficial and one is highly inaccurate; however, there is one that is outstanding in its depth and breadth. I certainly recommend that the interested reader locate a copy of *The Space Telescope* by Robert W. Smith.

CHAPTER ELEVEN

What Orion Really Looks Like

We saw in Chapter 8 that when the world's largest telescope, the 36-inch refractor at Lick Observatory, became operational in 1888 the Orion Nebula was and had been one of the first targets for powerful new telescopes. This remained true a century later when the Hubble Space Telescope became operational in 1990. What we found was truly amazing and has revised our understanding of star and planet formation.

Things were very different in 1990 than they were in 1888. In those earlier times it was simply accepted that the director and his key staff would select the objects to be studied first and any problems encountered with making the telescope operate as expected would be resolved on the mountain and outside of the scrutiny of a watching world. None of this applied when the Hubble was launched. Through television and the print media there was more attention being given to the Hubble than had ever been lavished on a scientific project. Astronomers aided and abetted this, often waxing hyperbolic with predictions of the revelations expected. I was probably as guilty as any and perhaps more than most, because by then I didn't have a major position of direct responsibility and was free to talk to the press.

We knew that interest would be great and that there would be an immediate demand for spectacular pictures. This demand set in process a program called Early Release Observations. Planning the early images that would immediately be released to the media and a waiting world proved to be a difficult process. It brought to the surface just how strongly the development-phase scientists felt about their programs. To understand this, you must first understand how the Hubble science programs were planned.

It was recognized from the beginning of the program that the scientists involved with the development of the Hubble would be the exclusive users in the early days of operation. This was generally accepted as an appropriate reward for having planned and built the observatory. Ultimately, it would be openly available to scientists who had no development responsibility, but had valuable scientific projects. The development-phase scientists were called the guaranteed time observers and the later observers the general observers. A product of this planned transition was that the guaranteed time observers were asked several years before launch to clearly formulate what their observing programs would be, thus allowing the future general observers to begin their own planning for other programs.

The guaranteed time observer programs included many of the most spectacular objects in the sky. Images of such objects would be good exemplars of the Hubble's power, and good candidates for the Early Release Observations program. Those images were to be released immediately, often before the scientist who had chosen those particular targets had had sufficient time to assess the data. Once the data was in the public domain, anyone could analyze it and perhaps rush to judgment, without the depth of knowledge of the guaranteed time observer. Even though the Early Release Observations of a guaranteed time observer's target were not accounted against their allocated reward in observing time, many observers were reluctant to see their targets included in the program. There was an underlying conflict of needs. We needed to make good images publicly available as soon as possible while honoring the rights of those who had invested more than a decade in building the Hubble and its instruments. The guaranteed time observer needed time to make observations and thoroughly analyze the data before going public with the result. This inherent conflict was compounded by poor communication between the office in the Space Telescope Science Institute, responsible for setting up the Early Release Observations program, and the guaranteed time observers themselves. This meant that there was an already-charged situation when the Hubble was launched.

Of course all plans for Early Release Observations went overboard when the optical problems with the primary mirror were discovered. The situation was chaotic, to say the least. The previously friendly media were now hostile to NASA and all involved with the Hubble project. Some of the most volatile scientists simply walked away from the project and their responsibilities, while others reacted by immediately going to work to get the best science possible out of the impaired observatory and planning for its repair. It was important to recognize that the nature of the optical problem meant that the core of the image of a star was almost as it should have been, however, this core contained but a fraction of the total light and the rest was spread out into an unexpected halo around that core. Very quickly a method of massaging the images was developed. This computer program took the test specimen images of known single stars and determined the exact shape of the blurry image. It then corrected the scientific images for this blurring, thus producing a final image that had removed much but not all of the effects of the optical problem. The original Early Release Observations program was scuttled and replaced with scientific instrument team–generated programs, which became the first useful images released to the public. This new program of early observations didn't account for all scientists who had been planning specific observations, often for many years. Eventually ruffled feathers were smoothed and the scientists who had legitimate scientific claims were involved with the observations and their use.

At the same time observations were planned to calibrate the cameras. In several cases, observations of a scientific target, including the Orion Nebula, could be made for these engineering purposes. Engineering observations were not required to go through the protocol of assessment for conflicts with guaranteed time observers' programs. You can imagine my surprise when I learned that engineering observations had been made with the primary imaging camera that were almost identical to ones in my own program, and that a scientific paper had been drafted based on those observations. Jim Westphal, the principal investigator for that camera, insisted that the observing team involve me once he discovered the conflict. However, it was too late to influence the paper very much. It was published in short order, despite a shortage of hard scientific information. Some of the claims made in that paper have not stood the test of time and subsequent observations. The calibration of the filters, which was the stated reason for making these engineering observations, was never finished nor published. Now that this "engineering" activity was public, I had a chance to participate in formulating the second set of observations in a different part of the Orion Nebula, one that would both satisfy the calibration needs and serve my ends, which were to characterize and explain the fine-scale structure of the Orion Nebula.

The second set of Orion observations were made in August 1991 with the original camera, followed by a further set of observations made just days after the installation of the new camera in December 1993. Since this new camera included optics that corrected for the problem with the primary mirror, it provided even better images than we had expected. The Early Release Observations confirmed the results of the second engineering observations, extended them, and were incorporated into a survey of the entire bright region of the nebula completed as part of my observing program.

Discoveries Made with the Hubble Space Telescope

These observations revealed not only a new layer of detail in the Orion Nebula, but also provided data for the discovery of fundamentally new processes. The most interesting of these has been the discovery of the class of objects now known as the proplyds. The proplyds are young stars with circumstellar clouds of gas and dust that are rendered visible by being in or close to an emission-line nebula. As I discussed in Chapter 9, we knew that such circumstellar clouds are a necessary part of the formation of new stars. We also knew that the Orion Nebula Cluster was very young. What we had not expected was that these circumstellar clouds would be rendered easily visible because of the Orion Nebula. With all the wisdom of hindsight, we now understand that this must be the case, and we should have been planning special observations just to detect the proplyds. Instead, the observations were made in order to study the fine-scale structure of the nebula and the discovery of the proplyds has become a good example of serendipity in science.

What we have found is well illustrated in Figure 11.1. Each of these objects was thought to be a simple star from ground-based observations. This is obviously not the case. Since this image was made with filters chosen to isolate the emission lines from atomic gas (hydrogen, oxygen, and nitrogen), it does not show stars well if there is a circumstellar cloud present. Only one of the four bright objects is a simple star.

The simplest object to understand is at the upper left of the figure. This is what we call a silhouette proplyd. It is a pre–Main Sequence young star surrounded by a circumstellar disk of gas and dust. As I noted in Chapter 8, the Orion Nebula is a thin irregular concave surface, glowing in the background behind the Orion Nebula Cluster. In the case of the silhouette proplyd, we are seeing the circumstellar disk in profile against this bright background. It is the dust that is making the disk obvious. At the

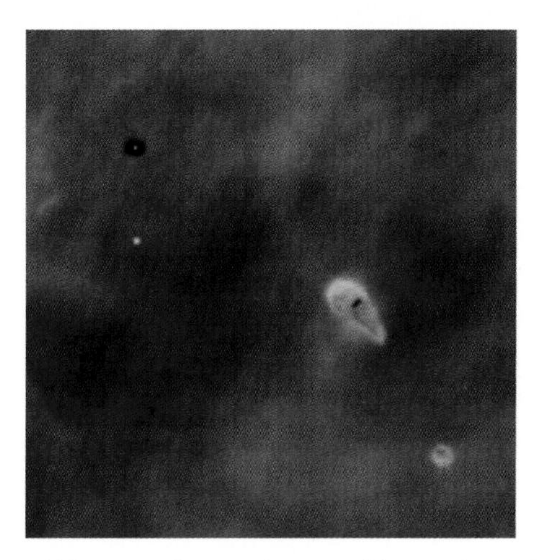

Figure 11.1 This early Hubble image of a central region of the Orion Nebula nicely shows that almost all of the objects thought from ground-based observations to be simple stars have turned out to be young stars with surrounding clouds of circumstellar material (proplyds). Sometimes we see the surrounding material as bright-rimmed, other times as dark objects, and sometimes as both (the author and NASA).

center is the young star, about one-third the Sun's mass. The disk itself is circular in form and we see it as an ellipse because it is tilted. Silhouette proplyds occur in a variety of forms, all of which can be explained as seeing circular disks randomly oriented with respect to the observer.

The bright object near the middle of Figure 11.1 is also a proplyd. In this case the disk is aligned almost exactly toward us, so that it appears as a thin dark streak. There is enough material in this edge-on disk to block out light from the central star. What is striking about this object is that it is surrounded by glowing gas. The process of fluorescence that causes the Orion Nebula to be bright is also causing the outer parts of the cloud of circumstellar material to be luminous. The bright side (the cusp) faces Theta1C, which is the source of the ultraviolet photons that produce the fluorescence. The portion to the lower right of the cusp is shadowed from direct illumination by Theta1C, and is illuminated only by the fainter radiation from Theta1C that is scattered by the nebula.

The object at the lower right corner of the image is a bright proplyd with a star too faint to show up in this image. We can see some indication of extinction by dust in the inner disk.

The basic difference between the bright and silhouette proplyds is where they are located. They are the same type of objects, but in different conditions of illumination. The silhouette proplyds are those young stars that are located within or beyond the obscuring veil of material that runs across the front of the Orion Nebula. The amount of material in the veil is enough to stop all of the high-energy ultraviolet photons from Theta1C. At the present we know of 16 pure silhouette proplyds, which is consistent with the foreground veil being about 3 light years in our direction from Theta1C. The bright proplyds fall in the enormous open volume between the veil and the main layer of the nebula shown in Figure 8.3. These objects are fluorescing on the side facing Theta1C, and whether or not we see the inner disk in extinction will depend on the orientation of that disk. Many of the proplyds had previously been recorded as simple stars, without comment. However, a study by Ed Churchwell at the University of Wisconsin with the VLA radio telescope showed that some were very peculiar. He anticipated the true nature of the bright proplyds.

Throughout the Hubble's Orion Nebula images we see many examples of shock waves. Shock waves form when a body is moving faster than the velocity of sound in the local medium. Sound waves are an example of the compression of air. This compression is periodic in the case of a sustained note from an instrument or voice, or a brief pulse in the case of something like the thunder generated by a flash of lightning. An airplane compresses the air ahead of it, but this compression is able to move out ahead of the aircraft as long as the plane is moving less than the speed of sound. However, when it exceeds the speed of sound the aircraft "catches up" with the compression wave that it forms, so that the air builds up in density in front of the plane. This standing wave of compressed gas is called a shock wave, so named because the conditions there are quite different from the surrounding gas. The density is always higher there and often the gas is hotter because some of the kinetic energy of motion has been imparted to the shocked gas. Shock waves can be formed when a solid body is moving supersonically, a supersonic jet of gas is moving into stationary gas, or if two streams of gas are colliding at a relative velocity that is supersonic.

Figure 11.2 shows a close-up view of the Trapezium region. In addition to the over-exposed images of the four brightest Trapezium stars we see a host of proplyds, each with a tail of material streaming behind it, pointing away from Theta1C. Several of the proplyds show a curved bright feature just in front of them and in the direction of Theta1C. These are shock waves, formed when a high-velocity wind rushing away from Theta1C collides with a slow wind of atoms coming off of the associated bright proplyd. These shocks are much hotter than the Orion Nebula. The green fan-shaped features at the 10:30 position from the brightest stars are artifacts from reflections inside the camera used for these images.

We saw in Chapter 9 that when a cloud collapses into a nascent star, there is a phase in which its material is drawn into the inner circumstellar disk. Magnetic fields permeate the clouds of the interstellar medium and these must survive in the collapsing protostars. The combination of magnetic fields and in-flowing material forms bipolar jets of material that spew out along the axis of rotation of the star-disk system. We see the same thing in the Orion Nebula. Figure 11.3 shows the bright,

Figure 11.2 This full-resolution close-up view taken by Hubble of the Trapezium stars shows how the proplyds often have tails that always point away from the brightest and hottest star in the Trapezium, Theta1C. That star dominates the radiation field, drives off material from the proplyds, and causes the region behind the proplyds to be in shadow (John Bally and NASA).

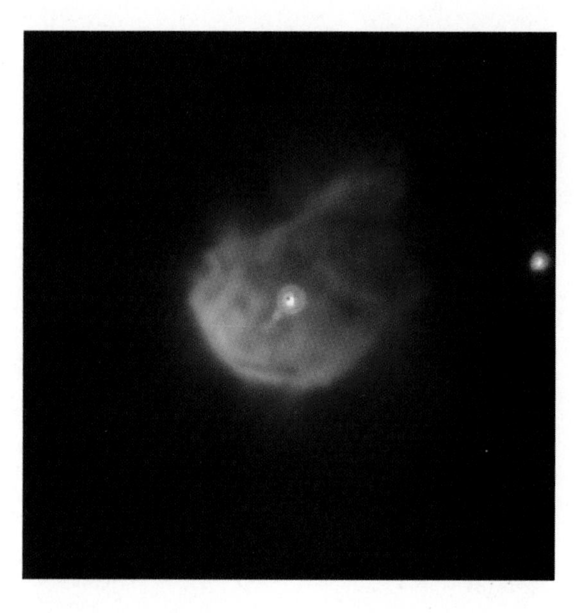

Figure 11.3 This, the largest of the proplyds, fluoresces ultraviolet light from a hot star on the edge of the nebula. Like many other proplyds, it has been recorded to have bipolar flow in a jet, although in this case the side of the jet coming toward us is much brighter than its counterpart (the author and NASA).

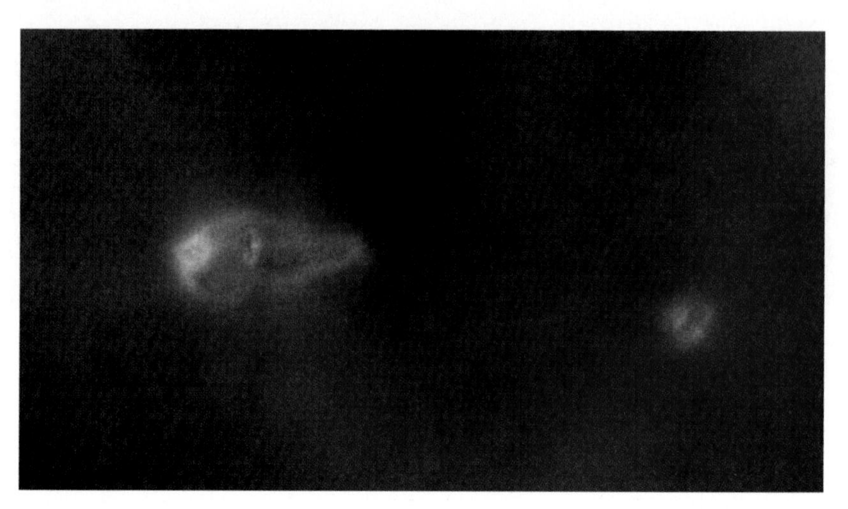

Figure 11.4 This close-up of the largest proplyd in Figure 11.1 was taken with a special set of filters. The blue light indicates light from neutral oxygen atoms. The blue glow surrounding the central disk is caused by the OH molecule. These molecules have been broken up into independent oxygen and hydrogen atoms by the starlight from Theta1C (John Bally, the author, and NASA).

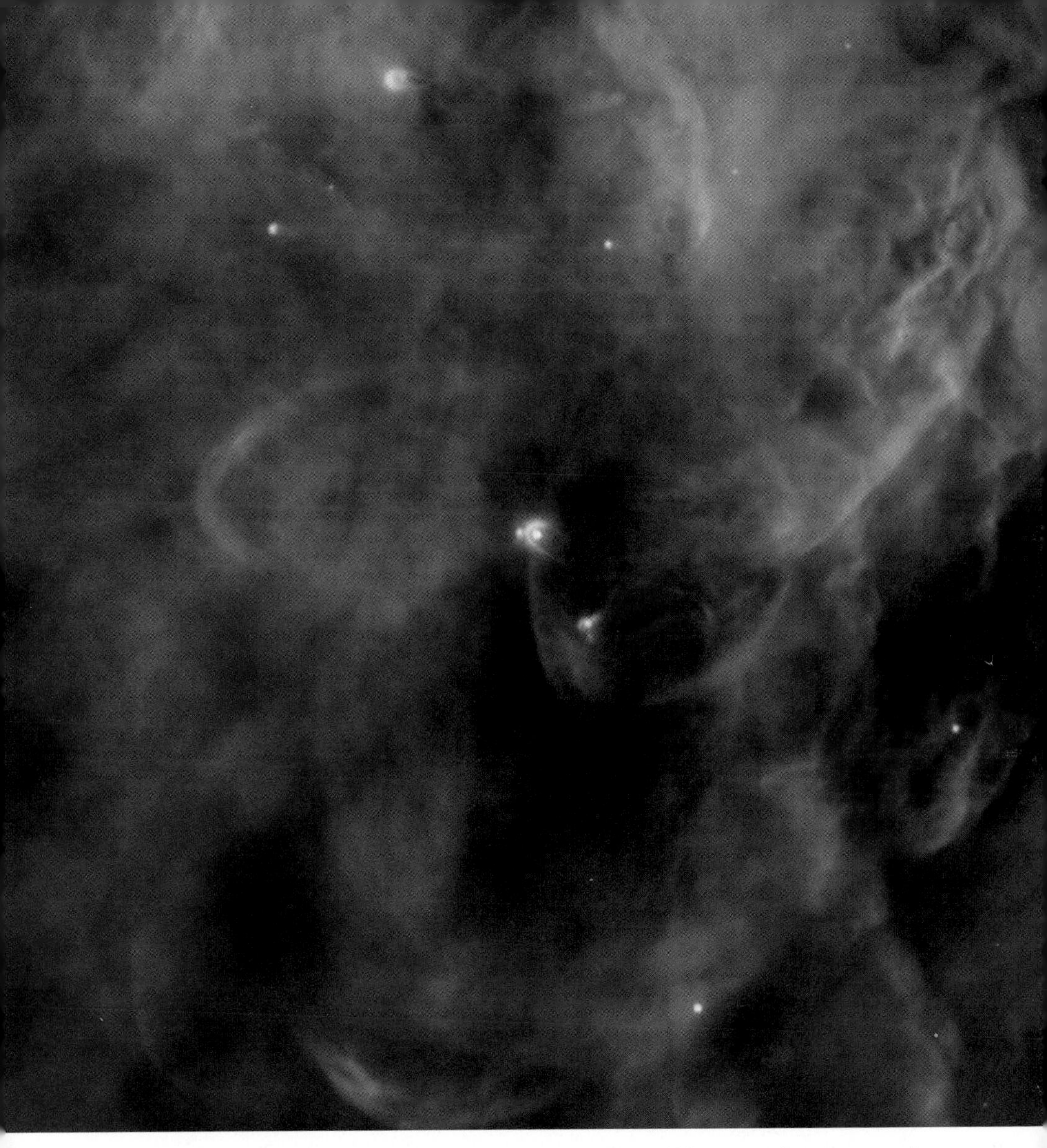

Figure 11.5 Complexity of the inner Orion Nebula is illustrated in this Hubble image. There are several proplyds—the brightest is in the center of the image. The bright cusps around the proplyds are caused by the fluorescence of light from Theta1C. The ensemble of large blue loops progressing toward the lower left in this image are shock waves created when a hypersonic jet of material strikes the ambient low-density gas out in front of the Orion Nebula, which is in the background. A portion of this jet is seen as the irregular yellow filament headed toward the center of the lower boundary of this image. This irregular jet is seen only as it emerges from the molecular cloud that lies behind the nebula (the author and NASA).

blue-shifted jet coming out of the inner disk around the star in the middle of the largest of the bright proplyds. This image indicates a faint hint of a counter-jet in the opposite direction. This counter-jet is confirmed in spectra made with the Keck 10-m telescope, which shows this material to be red-shifted with a velocity of 60 km/sec. Figure 11.4 shows a different perspective of the proplyd near the middle of Figure 11.1. In this case the blue filter reveals very low energy oxygen emission. The inner disk glows in this color because the oxygen there has just been liberated from the molecule OH. Other infrared Hubble images confirm that the inner disk is composed of molecules and dust, with essentially all of the atoms being caught up into molecules because of the high densities and low temperatures that apply there (like the inside of a molecular cloud). We can see the faint bipolar jet coming out perpendicular to the inner disk. Where the leftmost jet passes through the cusp of the bright proplyd, it has punched a hole in that material, much as the Space Shuttle passes through a thin layer of clouds low in Earth's atmosphere.

We see jets in 25 of the 300 stars imaged in the inner part of the Orion Nebula Cluster. We do not see jets in all of the nearby stars because there is an observational selection effect—the bright background of the Orion Nebula can mask over the image of a faint jet. There is another effect we've learned about from studying other star formation regions, where it is easier to see the jets. In those regions the jet outflow exists only for a small fraction of the lifetime of the disk. Jets are thought to be present only when material is flowing from the outer onto the inner parts of the disk.

These jets of material are flowing at a velocity of Mach 10—about ten times the speed of sound in the local gas. This means that if the jets collide with the nearly stationary low-density gas streaming off the front of the nebula, a shock wave can be formed. These are analogous to the shock waves that form in front of supersonic aircraft. We see the result in Figure 11.5, where a host of shocks being driven by jets is seen. Most of the jets have faded to invisibility against the nebula, but the hot gas in their shocks is quite visible. The fact that there are several cases of shock waves in alignment means that flow in the jet is not continuous, but must occur in starts and stops.

C H A P T E R   T W E L V E

What Is Happening in the Orion Nebula?

THE region around the Orion Nebula is a stellar nursery and one for which we have a bird's-eye view. Because of our perspective we can hope to understand both the individual and group properties of its population. This cluster of about 3,500 stars sits on the front of a giant molecular cloud of dust and gas designated the Orion Molecular Cloud-1. The process of forming these stars has either used up or dissipated the leftover material. We know this because the cluster sits in a rather open region, which allows us to see the stars individually. There is only the obscuration of the thin foreground veil and the circumstellar clouds that surround many of the stars. There are a few closer regions of star formation, but these are deeply imbedded in their natal material and none have the bright massive stars that characterize Orion.

Consider what it would look like to be within the Orion Nebula Cluster. The night sky would be truly spectacular. The night would be dominated by Theta1C, which would be shining ten times brighter than our full Moon, and the other Trapezium stars would also be brighter than the full Moon. There would be several hundred more stars that are at least as bright as the planet Venus at its most brilliant, while the other cluster stars would all be visible. Theta1C and the other Trapezium stars would appear blue-white and other stars would progress across the spectrum, with the faintest cluster members being a ruby red. Across one side of the sky would appear the Orion Nebula itself, looking much as it does in Figure 12.1, a glowing wall of gas. We would be able to see the proto-

Figure 12.1 The Hubble has allowed an unprecedented view of the Orion Nebula and its associated young cluster of stars. This mosaic was formed from 500 individual visual light images carefully spliced together. The colors come from isolating the light from different atoms and are balanced to what the human eye would see if near the nebula. Because the filters used don't include much starlight, the stars appear fainter relative to the nebula than they do to the eye (the author and NASA).

planetary disks of material around young stars, with the closer ones being several times larger than our moon appears to us. Those proplyds placed where the nebula shines would block out moon-size bites of the view of the nebula. From many of the stars we'd see jets of gas streaming out in arcs crossing much of the sky. It would be a spectacular place. Even the computer depiction in Figure 8.3 (made for an observer outside of the core of the cluster) only begins to approximate the spectacular view.

How Long Has This Been Going On?

I explained in Chapter 9 that it is possible to determine the age of a cluster of young stars. This is because the more massive protostellar clouds collapse more rapidly to the Main Sequence than their less massive companions. Stars of about the Sun's mass take several tens of millions of years to reach the Main Sequence, which means that if the cluster is quite young, stars still in contraction will be visible. This is the case for the Orion Nebula Cluster. Only the stars with masses about 2.5 times the Sun are on the Main Sequence. The rest lie in a wide band higher and higher above it as we look at cooler stars.

A detailed study of the distribution of these stars in the H-R diagram shows that star formation slowly began about 4 million years ago. Since then the rate at which stars formed steadily increased, reaching a peak about 300,000 years ago. In contrast with the 10 billion-year age of the Milky Way Galaxy, all these stars were formed just yesterday. The number of stars per unit volume increases dramatically as one nears the Trapezium, where the density becomes 20,000 times that of the stars in the Sun's neighborhood. Whereas stars in our neighborhood have average separations of about three light years, a star in the core of the Orion Nebula Cluster is only 0.12 light years from its neighbor. The average age of the stars decreases as you approach the center of the cluster, which indicates that the gas from which the stars formed is becoming more compressed with time. It is therefore not surprising that the most massive stars are found in the very center of the cluster. Since massive protostars collapse very rapidly, this indicates that either the conditions to form them occurred quite late or that it takes a longer period to collect enough material to them to form. We can't date the massive Trapezium stars because they are old enough that they've reached the Main Sequence yet are not so old that they have burned up enough hydrogen fuel to begin evolving off the Main Sequence. As we'll see in the next chapter, there are arguments that Theta1C is very young.

Brown Dwarfs and Rogues

I noted in Chapter 9 that most stars are lower in mass than our Sun and massive stars are quite rare. It is most difficult to determine the distribution of the lowest-mass stars, where the stars are intrinsically very faint. Orion is a particularly good place for resolving this problem since all the stars are visible in front of the molecular cloud. There are few unrelated foreground stars between us and the nebula. Almost all the stars we see in this direction belong to the cluster. We find that the number of stars with different masses is very similar to that of random stars near our Sun. This implies that the much older stars that we find nearby were formed in regions that obeyed the same sets of behavioral rules that we find today in Orion.

Where Orion is uniquely useful is in the study of the really low-mass stars. Brown dwarfs, stars just below 8 percent of the solar mass, will never become hot enough to burn their hydrogen fuel and become ordinary stars. All of their energy derives from gravitational contraction, except for a brief period in which they burn their small supply of deuterium. The majority of all known brown dwarfs reside in the Orion Nebula Cluster. This is not because they are so numerous there, but because that is where they are easiest to identify.

Below 1.3 percent solar mass even deuterium is not burned, and such an object can be called a planet. A careful examination of the coolest objects in the Orion Nebula Cluster indicates that there are about ten planet-size bodies, unattached to an ordinary star. Recall that our largest planet, Jupiter, has a mass of just 0.1 percent that of the Sun.

Over the last decade more than 100 planets have been detected in orbit about 88 stars near our Sun. How have these been discovered? The periodic Doppler shift of the host star lends a clue. The position of the host star is gravitationally perturbed as the stars swing about in their orbits.

The great surprise has been that most of these large planets are in orbits very close to their stars. This location and size makes them easy to detect: in fact, one was seen crossing the face of a star. It is argued theoretically that they must have formed farther out in the protoplanetary disk and migrated inward. The mass of these planets is comparable to the mass of the rogue planets—planets that roam freely, unattached to any single star—found in Orion.

The discovery of rogue planets poses an interesting question in semantics. Do they really deserve to be called planets? That name assumes both that the object does not burn nuclear fuel and also that it was formed in a protoplanetary disk around an ordinary star. A purist would insist on a unique name, such as planetars.

Regardless of the question of their name, how were they formed? Perhaps they formed as individual clouds that simply didn't trap enough material. Possibly they are planets that were once attached to a star, but were stripped away by gravitational collisions with other stars in this highly dense cluster. The answer is clearly that they formed as individual objects, because most of the planetars have a surrounding disk of dust and gas, just like the protoplanetary disks that surround the true stars in Orion. It is highly likely that these disks could have been pulled off along with the planetar if they are the result of gravitational collisions.

A City with More Than One Neighborhood

So far we have been talking about the Orion Nebula Cluster, its stars, proplyds, shocks, and assorted miscellaneous population. This certainly is the most important region of young star formation in Orion, but not the only one. Lying within the dark confines of the background Orion Molecular Cloud are two more groups of young stars. They are impossible to see in visual light because they are located behind the concentrated wall of dust and gas that lies immediately beyond the bright surface of the Orion Nebula. However, when you use sufficiently long infrared light, you can see what is there. Infrared light's wavelengths are large compared with the size of the dust particles, so that infrared light simply goes past most of the particles without being absorbed or scattered.

What you see in the infrared is nicely illustrated in Figure 12.2. In this case red depicts emission by molecular hydrogen (two hydrogen atoms bound together, much like the nitrogen and oxygen that constitutes almost all of Earth's atmosphere). This material is so fragile that it can be broken apart easily by radiation, so that it must lie inside the dark molecular cloud. The green component of the image is emission from iron ions of a type that can exist in the layer immediately beyond the Orion Nebula, but still within the cloud. Blue in this case is still an infrared wavelength, one that captures radiation from particularly cool stars. The Trapezium stars are so bright that they are overexposed near the center of the image. Up and to the right of those stars there is a concentration of molecular hydrogen emission. This region, the BN-KL region, receives its name from the two most intense infrared sources lying at the center. Eric Becklin, Gerry Neugebauer, Douglas Kleinmann, and Frank Low, for whom the region is named, were all pioneers in the study of the infrared sky. A close examination shows that to the north of the BN-KL region there are a number of red features, like fingers on a hand, and at the ends of these "fingers" there are bright green tips. Time-lapse imaging with the Hubble over a period of six years has shown that these fingertips are expanding away from the BN-KL region at velocities of up to 300 km/sec. If one extrapolates backward in time, all of the moving fingertips seem to have arisen from the BN-KL region about 900 years ago. We don't know exactly what is causing them. They could be the result of a gaseous wind coming off a hot star in the middle of the BN-KL region, or they could be a host of "bullets" fired off simultaneously as a group of young stars was created.

The center of the BN-KL region is best shown in Figure 12.3, which is again an image in molecular hydrogen. This time, however, the stellar background has been subtracted. This is one of the first images made with the Japanese 8.2-m Subaru telescope, so some flaws are evident. Nevertheless, it gives a remarkably clear depiction of how everything seems to be centered on one particular spot in the BN-KL region. This spot is an intense radio and infrared star known as IRc-2. It is probably a star about as bright and hot as one of the lesser Trapezium stars. In this case the star's radiation is absorbed by the dust in the surrounding

Figure 12.2 This infrared image was the first to reveal the clear nature of the fingers of molecular hydrogen spreading northward from the BN-KL region, which is the bright red area to the upper right (northwest) of the Trapezium stars. It differs from the visual wavelength image in Figure 12.1 because the infrared is dominated by emission coming from behind the material that produces the visual nebula (David Allen and Michael Burton, Anglo-Australian Observatory).

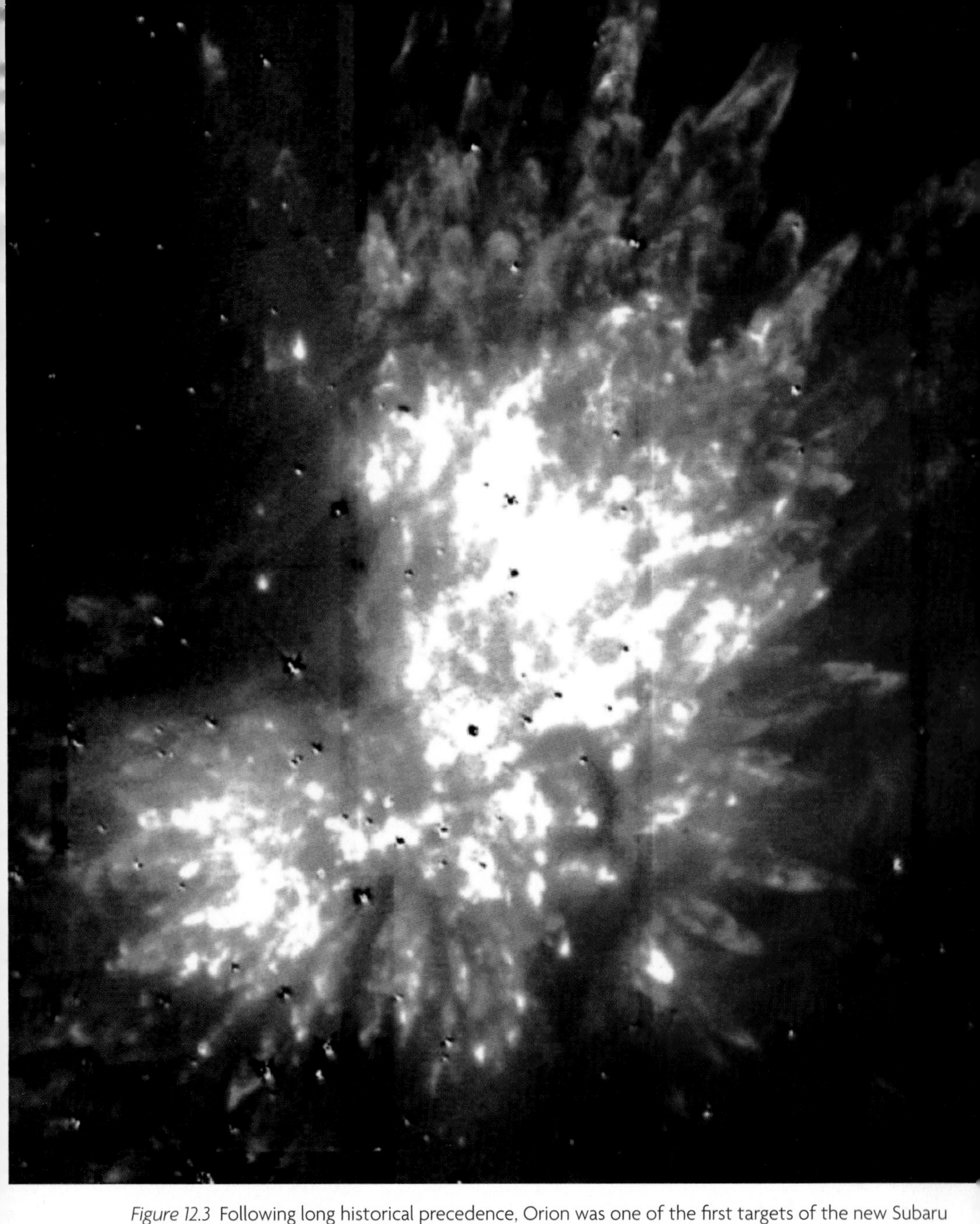

Figure 12.3 Following long historical precedence, Orion was one of the first targets of the new Subaru telescope operated by the Japanese on Mauna Kea in Hawaii. In this early image of the central part of the BN-KL region the bright emission is a result of molecular hydrogen and the strange small images are caused by the imperfect subtraction of the image of stars. The long faint straight lines are where the image was formed from multiple smaller images. One can easily see that the outflowing fingers and filaments all seem to have a common point of origin (Subaru Telescope, National Astronomical Observatory of Japan).

molecular cloud. This massive star is accompanied by a small cluster of lower-mass stars. The entire group is much less numerous than the Orion Nebula Cluster. We have no accurate knowledge of its age except that it must be less than about 5 million years old, otherwise the dominant star would have used up all of its hydrogen fuel.

The third center of recent star formation is a region called Orion-South. Its location is shown in Figure 12.4. Like the BN-KL region it too

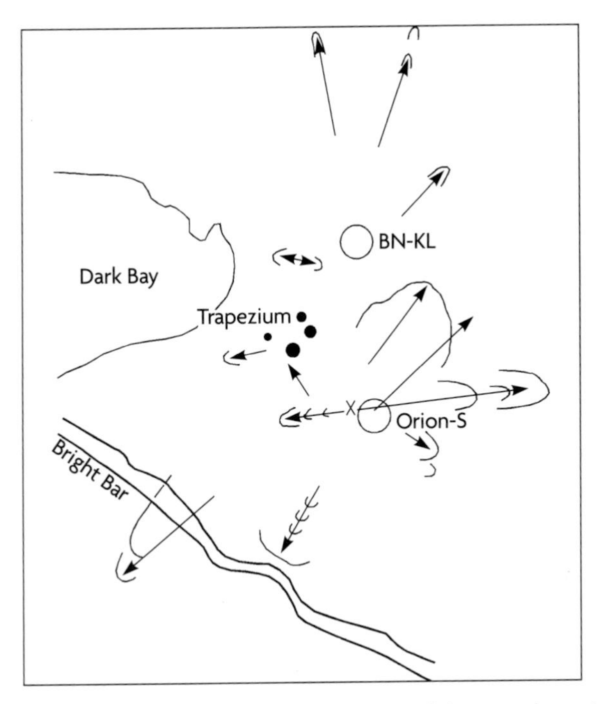

Figure 12.4 This cartoon covers essentially the same part of Orion as shown in Figures 12.1 and 12.2. The arrows indicate the motion of objects seen to be changing in position on Hubble images. The outflow originating from BN-KL is obvious, as are multiple features that originate from the Orion-S(outh) region. The *X* marks the position of the source of multiple outflows. At the present this source is unidentified at any wavelength of observation. Likewise, it is unclear how an object can produce multiple shocks. In the top of the center region there is an excellent example of a pair of shocks being driven by bipolar outflow from a proplyd.

is buried beneath the surface of the bright nebula. Infrared observations show that inside there is at least one star that is about 1 percent as bright as Theta1C. Radio observations can trace two bipolar outflows from this star. These are probably the result of a stellar wind from a hot young star pushing against the surrounding molecular gas, much like what is happening around BN-KL. Again, there is probably a small cluster of stars there. That cluster lies just beneath the visible nebula. We know that because we can see many shock waves and jets arising from multiple sources. Figure 12.4 shows that there are a large number of features appearing to move away from Orion-South. We can conclude that the driving source is located at the intersection of the backward extrapolation of these lines. The figure also shows that there are shock waves in front of a number of the Orion Nebula Cluster stars. These are shocks formed when a general wind of material blowing off of a young low-mass star collides with the low-density gas of the nebula itself.

In this chapter we have seen that the Orion Nebula is simply the most visible portion of a fantastic region of star formation. The associated cluster of stars is so rich that it contains really hot stars. Since hot stars emit most of their energy in the ultraviolet, this means that the nearby gas efficiently converts that invisible ultraviolet energy into optical light, which we see as the nebula. The cluster of stars contains stars of all possible masses, including large numbers of brown dwarfs that have only enough nuclear fuel to shine for a brief period and even planetars, which are objects so small that they will never become true stars. As if this was not enough, there are two additional regions of clustered star formation. One is to the northwest from the Trapezium stars, which seem to be the source of a general outflow of material, and the other lies just under the surface of the nebula to the southwest, producing beams of high velocity gas that break out into the nebula and form spectacular shocks. We seem to be lucky in our timing, because the nebula was probably not this spectacular in the past and won't be in the distant future.

CHAPTER THIRTEEN

Are We Alone?

MYSTICISM has been pushed back as our knowledge of what the universe is and how it operates has progressed. Modern people may stand in awe as they watch a total eclipse of the Sun, but would scoff at the idea of its being a harbinger of things to come. Epidemics such as the bubonic plague and our current pandemic of AIDS are understood to be the product of natural processes and not divine retribution. Thinking people around the globe have freed themselves from many of the superstitions that have inhibited improvement of our lot. However, our improved understanding of the universe has actually encouraged speculation in one area, the possibility of life elsewhere.

As long as the planets were points of light in orbit about Earth, the question of life beyond Earth simply didn't arise. However, with the change of perspective produced by the acceptance of Nicolaus Copernicus' Sun-centered model of the universe, which was published in 1543, it was understood that the planets were probably other bodies such as our Earth. If other planets exist around our Sun, why should they not accompany the other stars? If there were planets near other stars, then would they have life? Although the former Dominican friar Giordano Bruno was burned at the stake in 1600 primarily for his adherence to the hermetic religious tradition, one of his heretical positions was that life was common in an infinite universe and would exist on other planets. He speculated on this issue in the face of dogma and paid a terrible price. Had he been born a century earlier he might have speculated about life upon other planets without such dire consequences. At the end of the sixteenth century he could not establish that planets near other stars actually exist, nor that life might exist on those planets. He could only speculate on the reasonableness of assuming that extra-solar planets exist.

The existence of planets seemed more reasonable with the work of Pierre Simon de Laplace at the end of the eighteenth century. By then Newton's laws of mechanics had been fully accepted and their ability to explain phenomena here on Earth and in the sky tested many times. Likewise, the spiral nebulae had been detected visually with the most powerful telescopes of the day (recall that they were not established as being enormous star systems until early in the twentieth century). Laplace argued that as stars formed they went through a phase where they were surrounded by disks of dust and gas. Planets could form within such disks. This model explained the structure of the spiral nebulae and the confinement of the planets in our solar system to a thin plane. He was wrong about the former, but two centuries ahead of his time on the latter. We now have a more refined model of star and planet formation, but it is one that Laplace would recognize as simply a refinement of his own.

We now understand that the key ingredient in the processes of forming a circumstellar disk of material is the property called angular momentum. Although the linear momentum of a shrinking body remains constant and the body continues to move at a constant velocity, angular momentum makes a shrinking body spin faster. Recall that in Chapter 9 I described how stars form from fragments of clouds of gas and dust, collapsing under the force of their own gravity until they liberate enough energy through nuclear fuel–burning to arrest the collapse. Since the initial fragment will have some amount of angular momentum (through the cloud fragments bumping into one another), the protostars will have a certain amount of angular momentum. As a result, the cloud will spin faster and faster as it becomes smaller and smaller. A spinning body will produce an outward force, familiar to anyone who has ridden on a rapidly moving merry-go-round. If the outward force is equal in magnitude to the inward gravitational force of the remainder of the protostar, then that parcel of material will be left behind in orbit. Since the material on the equator of the spinning protostar will be going fastest, this equilibration of forces occurs there first and material is left behind in a thin plane. This material left behind eventually contains most of the angular momentum of the protostar. Evidence is provided in our own solar system, which formed from a protosolar disk: the planets contain almost all of the angular momentum in our solar system, yet little of its mass. If it didn't happen this way a star would not contract enough to burn nuclear fuel. Disk formation is a necessary step in the formation of a star.

Figure 13.1 shows what we expect a very young star to look like. The star will have a surrounding disk of material. Most of this will be confined to a thin plane, but it will have some material even high above and below that plane. This material is the building blocks for planets. Without such a disk the planets cannot form in orbits about a star. Our solar system must have looked like this at about an age of 1 million

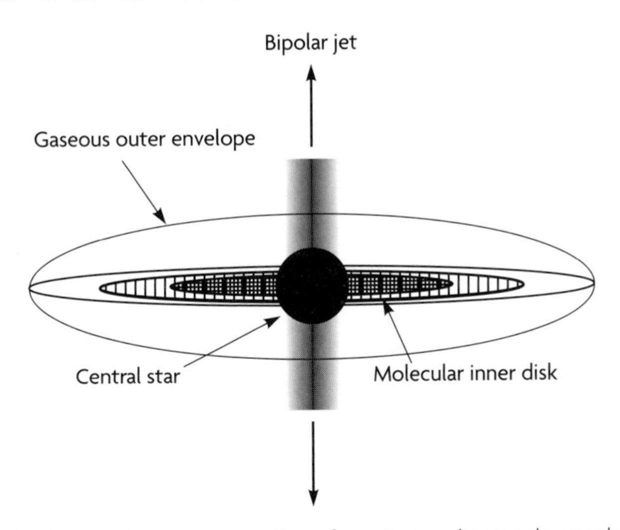

Figure 13.1 This drawing shows a cross-section of a protostar when nearly complete. The nascent star is surrounded by a disk of gas and dust. The dense inner parts of the disk are cool and the atoms are bound up in molecules. Molecules are not found in the outer disk where the atoms are unbound. The entire disk rotates about the spin axis of the central star. For a brief period during the shrinking of the star, a bipolar flow of hypersonic gas occurs.

years, which is just about the present age of the stars in the Orion Nebula Cluster.

Planets around Other Stars

The best knowledge of this subject comes from studying our own solar system. A most interesting division of types of planets is found. In the inner solar system there are four rocky planets (Mercury, Venus, Earth, and Mars) of similar size and density. In the outer solar system there are four giant planets (Jupiter, Saturn, Uranus, and Neptune) all primarily composed of the gas that dominates in the interstellar medium. I don't list Pluto as a planet because it is most likely a moon that has escaped from one of the outer planets. This division into two types of planets must reflect the very different conditions that applied in the inner and outer portions of the protoplanetary disk surrounding the young Sun.

The inner disk would have been much hotter. In this case the gaseous component would have evaporated as a rocky core of the protoplanets built up from the solid interstellar grains that would be a part of the disk. The two components (gas and dust) of the disk would be winnowed—exactly what we find in our solar system. The inner planets are smaller and rocky, having been formed primarily from dust, the less-abundant component (only about 1 percent) of the interstellar medium.

The outer disk would have been quite cold, so that gases would not have evaporated. This led to the creation of massive planets. (Jupiter, for example, is mostly composed of hydrogen and is 318 times as massive as Earth.)

By now, a hundred or more planets have been detected in orbit around other stars. They seem to be quite different from those in our solar system. In many cases giant planets like Jupiter are found close to the host star. Nevertheless our solar system's structure is not an anomaly—it is likely that these other planetary systems are instead peculiar. This argument is based on the method of search.

We don't see these other planets directly. We must look for the wobble in the velocity of their host stars as the planets move in orbit around them. Both the star and the planet move around a point called the center of gravity. This is like a seesaw. The massive object (the star) sits very close to the center of gravity, whereas the much less massive planet sits far away. Their orbital velocities around this center of gravity depend upon their distance from that center. This means that although the planets may be moving at velocities of tens of kilometers per second, the stars have a vastly smaller velocity. The closer the massive planet is to the host star, the faster the velocity of the star. A careful study of the radial velocity of the star using the Doppler effect—measuring the shifting of the lines in the spectrum of the star—will show a periodic variation of the star's velocity. The period will be equal to that of the time it takes the giant planet to go around in its orbit. Searches for these periodic variations in velocities were successful once the velocity could be measured accurately to within a few meters per second. If the other stars and planets had been like our solar system, this accuracy would not have been enough. It is remarkable that we can now measure a star's velocity with an accuracy that equals the speed a human can run!

When the changes in the velocity are analyzed we find that these other star-planet systems are quite different from our solar system, consisting of giant planets close to their host stars—so close that they must be spiraling into the star. The time they will spend before being absorbed by the star is but a brief interlude in the star's long life span. This incongruous conclusion is the result of the observational selection effects that are operating in the search. The radial velocity searches can only find anomalous star-planet systems, because only systems like this (a giant planet close to the host star) will have detectable velocity changes over short periods. Scientists are therefore finding the types of objects that are the goals of their search. This is not to denigrate the importance of these studies. They clearly establish that planet formation has occurred elsewhere, and may be a common event. Unfortunately these searches don't tell us what fraction of stars have planets, only that planets exist around some other stars.

Planet Formation in Orion

Orion is far enough away that stars in it that resemble the Sun are too faint for the velocity analysis that has led to the discovery of giant planets around some nearby stars. However, we can determine if the building blocks are there. By building blocks I mean the dense circumstellar disks that would allow the creation of future planets. As discussed in Chapter 11, Hubble images of the Orion Nebula revealed the new class of objects called proplyds. These are rather ordinary young stars with surrounding disks. What makes them special is their close proximity to Theta1C and the Orion Nebula, which makes the conditions for discovery and investigation of them quite different from young stars in other regions. There is no reason to expect that they are intrinsically different from other young stars, but they are all just about the same age and distance from us, which is a particular advantage in their study. We now know that the inner high-density portions of these disks are molecular, indicating that they are cool, which could allow the formation of even giant planets. We can't prove that the innermost parts aren't too hot to prevent giant planet formation because Jupiter's distance from the Sun would only be about one-fifth of the angular resolution provided by the Hubble. Resolving that is a challenge for future telescopes. What study of the proplyds in Orion has revealed is that essentially all of the stars there have associated circumstellar disks, which means that all of them have the building blocks of planets. Rich clusters like the Orion Nebula Cluster are thought to be the prototypes for formation of most stars. What goes on there is probably what has gone on in most other stars.

The Probability of Planet Formation

We know a surprising amount about how planets have formed in our solar system. This is largely because we have a good idea of the composition and interior structure of all the planets through observations and samples. In addition, meteorites are continuously impacting Earth, bringing samples of material that are available for intensive laboratory analysis. Here's the commonly accepted scenario: The interior planets, like Earth, formed in a short period of time, possibly less than 100,000 years, from material depleted in the light elements like hydrogen. In contrast, the outer planets, like Jupiter, formed over a period of millions or tens of millions of years through a process of first forming small bodies (planetesimals) that then coalesce to form giant planets, having retained their original composition from before the star began to form. Lacking any good reason to think other stars and disks would vary from this pattern, we have generalized these processes to other stars and protoplanetary disks.

There is something very dramatic happening in the Orion Nebula Cluster proplyds that affects how many of those stars will actually form planets and what kind of planets those would be. Recall from Chapter 8 that the nebula is actually a thin layer of fluorescing gas on the surface of the background molecular cloud. The glowing gas is at a much higher pressure than the nearby part of the molecular cloud so that it expands away from the cloud. The same process is going on in Orion's bright-rimmed proplyds because their surfaces too are heated up, creating an over-pressure situation. I've used spectrographs on the Keck Telescope and the Hubble to directly measure how rapidly this material is "boiling off" from the proplyds. Both measurements are in agreement, with the Hubble determining the value more accurately as a millionth of a solar mass of material per year. If the proplyds have disks that are like nearby young stars, they will have about one-tenth of a solar mass of material. Combining these two numbers means that the disk would be completely "boiled away" in 100,000 years.

This disk survival time presents a conundrum. The average age of the stars in the Orion Nebula Cluster is about 500,000 years. On the other hand the disks should survive for only 100,000 years, yet there is no evidence that they have been depleted. The proplyd disks should have dissipated within one-fifth of the lifetime of the cluster, yet there they are in 85 percent of the stars. The resolution of this riddle probably lies in the fact that the star that causes the nebula and proplyds to fluoresce must be younger than the other stars. The formation of a hot luminous star like Theta1C will heat up all the surrounding gas and stop the further formation of stars.

In other words, the proplyds near Theta1C will have only a limited time to form planets before their disks are dissipated. If the solar system really is a good model for planet formation, then this short lifetime means that only low-mass planets will have had time to form. The building blocks will have been destroyed before a giant planet has time to form.

Does this mean that no stars in the Orion Nebula Cluster can eventually form giant planets? No. Recall that the proplyds are seen with a range of chiaroscuro, as seen in Figure 13.2. Some are bright-rimmed, some are bright-rimmed with obscured portions in the middle, and some are entirely dark and seen only against the background glow of the nebula. The reason for this is that those silhouette proplyds are located within the foreground veil. This neutral material at the front shields these proplyds from the highest-energy photons coming from Theta1C.

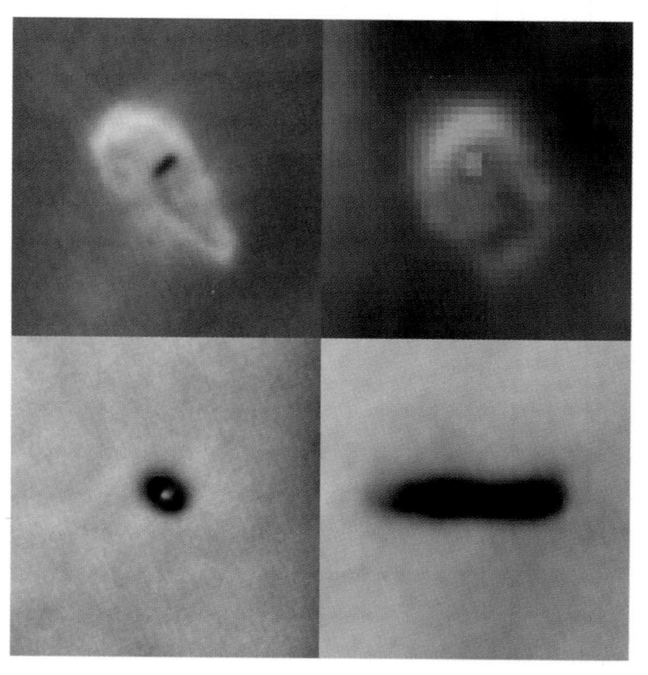

Figure 13.2 This mosaic of Hubble images illustrates the variety of the proplyds in the Orion Nebula Cluster. The lower objects are in the foreground and are seen in silhouette against the bright background of the nebula. The nebula's light is obscured by the dust component of the circumstellar disks. The bottom right disk is seen almost edge-on whereas that on the bottom left is viewed at an oblique angle. The top two proplyds are both close to Theta1C and that star's radiation is fluoresced by gas in the outer parts of the circumstellar clouds. The disk of the proplyd in the upper left is so nearly edge-on that there is no direct evidence of the central star, whereas the proplyd on the upper right is viewed from a higher angle and the star is obvious (the author and NASA).

These objects aren't being heated up by the bulk of the radiation from Theta1C and will evaporate material much more slowly. About 10 percent of the members of the cluster are thus shielded and remain candidates for forming both terrestrial and giant planets. If we can generalize, this means that terrestrial planets may indeed be common, but giant planets are more rare. Planets like Earth may well exist in Orion. However, having such planets doesn't establish that life exists there. Almost certainly it does not, since the Orion stars are so young.

Do We Have Cousins?

I began this chapter by remarking upon how increased knowledge has tended to quench speculation with the exception of one subject area. Our newly found appreciation that planets are possibly ubiquitous makes it more tempting to ponder the existence of other life in the universe. Life on Earth seems to have started early in its history, at only a fraction of the current age. This occurred soon after the surface became well formed and water began to cover most of Earth. The majority of the time that life has existed here, the life forms were very elementary. Once self-replicating multi-celled structures began to form, the process of evolution began to accelerate. At that point it was almost inevitable that environment-controlling creatures (us and our successors) would be created. Even if those first life-forms had been created a few billion years later or the watershed epoch of multi-celled organisms had come later, the process would have been the same. This is not to say that it was inevitable that sentient bipedal species with twenty digits would evolve, but merely that the processes of selective evolution will always lead in the same direction, given enough time.

This process can only occur within a limited range of conditions. There must be a large number of atoms and molecules close together in conditions of temperature and density that allow complex molecules to build up, then retain their structure for long enough to begin to replicate themselves. Giant molecular clouds form complex molecules on the surface of the grains in their middle, but these clouds have short lifetimes that destroy the special conditions before the process can go very far. This is because their temperatures are so low. The atmospheres of cool giant stars allow the capture of most of the heavier elements into molecules, but the temperatures are too high to allow large molecules to form. The best conditions for life are found on planets.

The orbits of most planets are very stable, which means that their average temperature will be almost constant throughout the main sequence lifetime of their parent star. They have the potential for forming a solid nucleus, which means that most of the material will not be circulated into extreme conditions, and a sufficiently large planet will have a gravitational field strong enough to retain an atmosphere. This means that identifying potential hosts for life lies within the domain of the astronomer.

About forty years ago the astronomer Frank Drake formulated the equation bearing his name. The Drake equation gives the probability of life occurring elsewhere in the Milky Way Galaxy that is able to communicate with us. It is a simple equation—a series of numbers multiplied by one another that represent increasingly restrictive subsets, beginning with the rate of formation of suitable stars and ending with the fraction of civilizations on viable planets circling such stars that have attempted extraplanetary communication, and the length of time such civilizations have attempted such contact. The basic approach of the equation is to reduce each factor of the equation to an independent, quantifiable multiplier.

This approach is like calculating the number of red-haired women within a specific acre of land in the United States at a particular time. Taking the simplest approach, this probability would be the product of multiplying many factors. First, how many people are in the United States at that time? Next, what fraction of the population is female? What is the fraction of all females who have red hair? Then we would divide by the number of acres of land in the United States, and so on. Of course it is not really that simple. If we chose a Sunday afternoon in autumn and the acre of land is part of a professional football stadium, then the concentration of people is going to be much greater than the average population per acre. However, the number of females may be fewer than the national average. Then we must consider whether red hair is in fashion, and on and on. My point is that even the simple Drake equation could produce wildly inaccurate results unless we can account for all its complex factors.

The astronomer can contribute to the data for the Drake equation by quantifying its first few terms. Since the original formulation of the equation we have both redefined some of the terms and begun to get numbers for others. The number of stars in our galaxy is becoming better defined at about 20 billion. The fraction of stars that have circumstellar disks, the potential building blocks for planets, is almost unity. The fraction of these that can form planets with solid cores is a subject of current research. We see in the Orion Nebula Cluster that the circumstellar disks possibly may be destroyed through overheating from Theta1C. If our understanding of the timescale for this destruction and the theoretical predictions that terrestrial planets can form very rapidly are correct, then a significant fraction of all stars with disks can form terrestrial planets.

We then have the interesting challenge of determining how many of those terrestrial planets will form in the habitability zone. By this, I mean a zone around any one star where the temperature is acceptable for the existence of life. The fraction will certainly be greater than zero, because terrestrial planets form in a high-temperature portion of the circumstellar disk. The complexity of refining our definition of the habitability zone is well illustrated by our solar system. One definition of the habitability zone is that it can sustain liquid water. This makes sense considering its ubiquity and immobility of molecules in its solid state. In its gaseous state (steam), complex molecules would be quickly broken down. If Earth had no atmosphere, it would have an average temperature just at the freezing point of water, with temperatures in the liquid zone only found near the equator. But we do have an atmosphere, which has a greenhouse mechanism operating so that the average temperature is higher than otherwise expected. Venus is closer to the Sun than Earth and should therefore be a bit warmer, but the dominance of carbon dioxide in its atmosphere has created a runaway greenhouse effect. Venus's surface temperature is hot enough to melt lead. Mercury is so close to the Sun that liquid water cannot exist there, either. Mars has only a thin atmosphere at present that barely warms the planet. The atmosphere was thicker in earlier years, and there is a growing body of evidence that liquid water may have been present on the surface for some period (raising the possibility that we might one day find fossil evidence of extinct life there). Using our solar system as a guide, only our planet lies in the habitability zone, although under only slightly different conditions, three of the four terrestrial planets might have been suitable.

We astronomers know much more than we did when the Drake equation was first formulated, but we clearly don't know enough to nail down our parts (the first parts) of this equation. Evolutionary biologists are hard at work solving the middle parts, dealing with the origin and evolution of life. However, all scientists are mostly developing an appreciation for the complexity of the problem. Philosophers and social scientists may never be able to calculate the final terms of the equation, which deal with the intentions and life paths of civilizations yet unimagined. This does not mean that the question of other life in the universe must be dropped, for it is fundamental and important, although we need to address it patiently and with vigor. However, if forced to declare if I think there are other planets with life, I would bet there are. However, I don't expect to live long enough to collect on the bet.

CHAPTER FOURTEEN

Outsmarting the Fickle Goddess of Science

THIS BOOK is on the surface a collection of facts concerning what we know about the Orion Nebula. To describe this particular nebula, however, I've explained the methods astronomers use to draw conclusions about objects in the sky, and the context of this knowledge. How does an understanding of the Orion Nebula fit into the bigger picture—an understanding of the universe itself? Although my subject in this book has been astronomical, the scientific process is essentially the same, whether it is geology, biology, or any other discipline worthy of being called science. We have heard different siren calls, but all of us who spend a lifetime in scientific research have been trying to outsmart the fickle goddess of science.

To understand the process of science you must appreciate the difference between information and knowledge. Put succinctly, information is a body of facts, whereas knowledge is the understanding of what those facts mean. We live in an era when it is easy to be inundated by facts. Televisions and computers are present in almost every home. Television is largely used for entertainment and consumer decision control, but computers provide access through the World Wide Web to more factual information. Computer information is filtered by the persons who have made that information available, of course, so that the ease of dissemination of information also means that it is easy to spread misinformation. What are now called urban legends have always existed. I would argue that they are but variations on the process that has given rise to creation myths in most societies. Whereas technologically primitive people evolved those myths slowly to represent their societal needs, the ability to rapidly share information has compressed the timescale for refinement and dissemination of these myths.

The information upon which science is built increases monotonically; that is, it always increases. The facts that we knew before will always remain. They may be interpreted in a different fashion, or may even be considered irrelevant, however, they will still be there. Of course if they are proven wrong, they disappear. New types of information are often created. The development of the astronomical telescope or the laboratory processes for sequencing DNA molecules have introduced types of information that had never been considered. The new information may seem vastly more useful and relevant than the earlier information, but that older information will always be there.

Knowledge is not monotonic. Sometimes we understand something incorrectly and accepting that it is wrong is often harder than the original acceptance. The usual path to understanding is that a set of facts slowly builds up. These can be independent but related. At some point an attempt at understanding is made. This understanding can be formulated as a law—"things always behave according to a set of rules"—or as a theory—"things happen in a certain way because of a specific effect." We can lump these two approaches into something called a model or paradigm. To be acceptable, a new model must be able to explain all of the previous factual information. If there is one credible and relevant fact that is in disagreement, then the new model must be wrong.

The Model of Our Solar System

Consider the historical case of the model for our solar system. The movement of the Sun, Moon, and planets across the celestial sphere had been charted for two millennia before Ptolemy formulated his Earth-centered model. This model was highly successful because it explained all of the known facts about the apparent motions of these celestial bodies. Over the next millennium, as more facts became known through more detailed and longer study of the motions, it was noted that some of these new facts didn't agree with the original model. However, they could be explained by altering the model slightly, by adding epicyles on epicyles or displacing the center of motion of some of the objects. The changes were viewed as necessary, but remained within the intellectual framework of the model. However, after enough of these additions, some astronomers admitted that something must be wrong.

The Sun-centered Copernican model for the solar system was attractive in its simplicity. Immediate resistance to this new theory was based not so much on facts, rather upon doctrine. The most public resistance centered around a literal interpretation of the Old Testament, stating that God had commanded the Sun to stand still. However, there were other, more quantitative objections. Copernicus's model predicted motions that were no closer to the observations than the Ptolemaic model. Why give up the old model for the new? The shortcomings of the Copernican model lay in the assumed circular orbits. Johannes Kepler demonstrated that if the orbits were ellipses, the model was functional. The correct information now existed, but it still had to be accepted. Newton, building on the Copernicus-Kepler model, came up with the Universal Law of Gravitation, explaining why those elliptical orbits existed, and subsequently describing the structure and motion of the stars. This Copernicus-Kepler-Newton model of the solar system proved absolutely correct until late in the nineteenth century, when it was seen that the orbit of Mercury was slightly different from what this model predicted. Unsuccessful attempts were made to save the model by arguing for unobserved inner planets that were perturbing Mercury's orbit. It was only when the Theory of General Relativity was developed that the peculiarity was understood. The model was still correct, but the Newtonian laws of mechanics were not the final word on how moving bodies behave.

The Model of the Orion Nebula

The intellectual history of the Orion Nebula is a good example of how science is done. Originally the Orion Nebula was simply an object painted on the sky. Creating a model of it was simply not a pressing issue. It was clear that it was composed of stars (the bright ones), but it was not clear what was producing the nebula. Smart money would have said that the nebulous part was a cloud because there was such a sharp contrast between the stars and the nebula. The alternative was that there were so many fainter stars that they simply appeared to blend together. Proof that the nebula was gaseous only came in the mid-nineteenth century when the new technique of spectroscopy was able to show that most of the light was in emission lines, just like the spectrum of a gas here on Earth.

Soon the new technique of photography could image fainter nebulae and perform spectroscopy on the brighter ones. It was seen that the gaseous nebulae (in contrast with the spiral nebulae, which turn out to be spiral galaxies) were both ubiquitous and possessed similar characteristics. This meant that the Orion Nebula was but one of many similar objects in our galaxy. Edwin Hubble's research showed that the type of star associated with a nebula is what determines whether it is an emission-line nebula or a reflection nebula. His work was strictly a straightforward interpretation of observational facts, but this bifurcation stimulated Stromgren to try to explain the nebulae in terms of physical processes. The result of that study was a comprehensive and general model for gaseous nebulae. They were clouds of interstellar gas and dust. Those with nearby hot stars were able to ionize the hydrogen gas and this gas would fluoresce the intense ultraviolet radiation of the stars into optical emission lines.

Stromgren's model for what others called Stromgren spheres was wonderfully successful. At a stroke it explained hundreds of objects dotting the night sky. It was almost too successful. As more quantitative observations of Orion became available through radio observations of its surface brightness (which were free of any effects of extinction by dust), a conundrum arose. If the nebula was spherically symmetric, as the Stromgren-sphere model assumes, then it must have the greatest density of the material nearest the dominant star, already recognized to be Theta1C. This could be viewed as only an additional feature of the Stromgren model, but such a concentration of material must have a much higher pressure than its surroundings. It would expand and dissipate in only a few tens of thousands of years. The problem became worse when a detailed analysis of the spectrum in those emission lines, which could identify the gas density, showed that the gas was much denser than indicated by the brightness. This made matters worse. The timescale problem was ignored, and the facts were all reconciled by assuming that the gas was actually compressed into dense clumps. The size and density of these clumps needed to be just right to explain the observations.

It took a nonspecialist to solve the problem. He understood the basic physics, but didn't have a personal investment in the spherically symmetric model. He noted a systematic difference in radial velocity of different ions in the nebula, meaning that we were likely looking at a wall of glowing gas, rather than a three-dimensional spherically symmetric cloud. Once the slab model was proposed, all the other facts fell into place. Clumps of varying size and density weren't necessary. Density and brightness were reconciled.

In the following two decades we have built up a more sophisticated model of the nebula, but additional data has not compromised it. The jewel in the crown has been the application of the superb imaging capability of the Hubble to the Orion Nebula. This has now shown that the cluster of stars associated with the nebula are cradled in the basin of the nebula. Outflows from these young stars are forming shock waves throughout. The Hubble images have also revealed circumstellar clouds of gas around most of the stars, providing confirmation of the link between the formation of stars and the high probability of the formation of planets. Throughout this entire process, theories and models have been built up and torn down, all in the service of seeking scientific truth.

I still study the Orion Nebula, always trying to adhere to the basic tenets of how science is done. Of course, this means I may need to accept new and contradictory evidence that could result in inaccuracies in many parts of this book. Nevertheless, today's tale is worth the telling.

Index